IMPRINT O

A

WORKS VI ARCHITECTURAL ASSOCIATION

FOREWORD

A young girl, jolted out of the familiarities of home; her awareness of place, substance, smells, noises, time, light and circumstance is heightened, the narratives of her life expanded.

Such is the story of *Imprint of India* … it is a primer of that heightening, that expansion … a primer of the impact of place on person.

It is published in parallel with the exhibition called CLIMATE REGISTER, it being realised that the messages of that exhibition need many channels if they are to penetrate.

Peter Smithson
October 1994

Cover design by Alison Smithson based on a photograph by J.S. Lewinski, 1972

Imprint of India has been published with its companion volume *Climate Register* to coincide with an exhibition of four projects by Alison and Peter Smithson held at the Architectural Association from 5 October to 4 November 1994. The publications have been produced through the AA Print Studio by Dennis Crompton (technical co-ordinator), Annie Bridges, Ann Cheatle, Jenny McIvor and Marilyn Sparrow.

Origination by Target Litho
Printed by E.G. Bond Ltd.

ISBN 1 870890 49 3

AA Publications, 36 Bedford Square, London WCIB 3ES

IMPRINT OF INDIA

Alison Smithson

PROLOGUE

RESPONSE TO CLIMATE, TO PLACE

This is evocation fiction about the British in India; the analogy is of a distantly observed firework display of independent and quickly passing sightings. The set-pieces – displays that touch on all the senses that compose experience – in juxtaposition convey a more extensive scene than any straight-through telling. The short scenes overlap, seem to contradict, relate apparently randomly. Characters appear, only to change names, their positions in life and in time. A scene re-occurs but is more confused; even more aggravating. The same scene can mutate… is it repeatedly experienced; or is it remembered? A mirror-image of a scene appears, yet in an altered context. Staccato phrases are the chemical powders of an underlying theme where girl and boy see, speak, wonder: the resultant flashes of shock of recognition are sparked again and again, for pleasure, for puzzlement; and the encounters flicker like another's fireworks in a night sky whose rapidly spent fragments, in their descent, interweave and interrelate in a variety of patterns. There is no one beginning, no single end; just the worrying over this magical happening through meeting.

IMPRINT OF INDIA: POINTS OF DEPARTURE

In the 1960s, the simple act of turning a street corner in Bombay might find you facing your own childhood… a Kodak Lady, in a blue-and-white-striped dress is holding a Box Brownie, a full-size figure, preserved since the 1930s in the stove enamel on the metal placard: such advertisements once stood outside every English seaside chemist and postcard shop of any pretension. Or, again 1960, rounding a corner in Bombay, you face a building so like the Mechanics Institute Bradford that, there on the hot pavement, you are impacted by two very different, evocative senses of one's own history. In Chandigarh, the question carefully put: "Will you have your eggs hard-boiled or soft?"… the aged outdoor servant has been brought into the new suburban detached house of a place that did not exist before the British left, only to ask this for the smart Independent-India servant. That they knew to ask! That they still care you should be asked! Edinburgh of the 1930s, and a handful of similar places in which lived professional people who had come home to retire, are hereby recalled. So many brought their curios home from service in bank, army, hospital, education or mission. The box rooms of such people held an assortment of ships trunks with overlaid labels, and business-like picnic baskets.

TRAIN MOVING 9

Great Indian Peninsular Railway, passing over Ghats, broad gauge, standard gauge for India, 5' 6". 2:8:0 goods loco, 2:8:4 heavy tank engine.

Central Indian Railway, Bombay – Bawda, metre gauge. 4:8:0 goods loco, India-built due to world shortage of engines around 1889.

South India Railway.

East India Railway.

Nizam's Guaranteed State Railway, Hyderabad–Bombay.

Madras Railway, 5' 6" gauge.

Trains with such names as "Delhi Mail"; "Bombay Calcutta Express". Colour distinction of Class standard throughout India, though shades varied. First Class white, Second Class green, Third Class gold. The train could be made up so that the servants' gold or ochre carriage was in communication with the First Class carriage.

BREATH OF INDIA 14

Part of the journey already consumed; and then by a sudden settling down of the engine to speeding home, the comfortable vibration rises and introduces itself into the memory base from which there start to rise such pleasant nostalgias as to cloak the new journey with an aura of great respectability and well-being.

WORK IN ALIEN CLIMATE 20

Certain make-shift arrangements characterise a period. In our time the oil drum has been the ubiquitous civic furniture of India... perforated, they protected trees, held work fires; filled, they delineated road improvements – roundabouts, widening; weighted, they supported signs, telephone poles, formed roadside seats. Flattened, they became sheetmetal fences, house sides, sheetmetal roofs; the round drum ends used as patches to cover holes in buildings, as metal disc tiles for roofs, or as wheels; guided by a hooked wire to a centre hole, play things to a running child.

MORNING TEA, AFTERNOON TEA 26

At any point in time during the period the British were in India, an individual, newly arriving, stepped into a strange world in which there nevertheless existed unexpected and poignant signs of home: statues and drinking troughs, names of buildings and streets, indication signs, buildings. This accumulation, the result of a myriad of tiny administrative acts, any individual could have found touching. But the placing, routing, dividing, in the town and the landscape, a trained eye could also read as a British presence, and by these indications maybe directly moved.

At any point in time during the British presence in India the whole place could have been found disturbing, unsettling to some minds when all the strange people, all the strange and familiar images and memories, rudely crowded the bloodstream and jostled for pride of precedence.

LAYERS OF OCCUPATION 32

To be woken early by unusual light and quiet; to step out into the cool air and survey the coming day: or at the end of a day, to be confronted by the closing colouration of the sky; to sit and give rest to body and anxiety. By contrast, clutched into the speed-squelch of big bus wheels on tarmacadamised road, continuous tune behind the immediate sounds of a confined group's reaction to a hot afternoon of travel: Indian curtained bus bowling along the Trunk Road engineered by the Moguls, refurbished

Chapter 1

TRAIN MOVING

Train moving… the compartment is a safe box in which to play house. Long periods are thus spent, alone.There is no communication with the other compartments. Descent from the train while it moves, impossible. No individual, however exalted his post, can oblige the engine driver to stop, disobey the regulations governing the railway system – unless he makes himself liable to a penalty of eight rupees. The short train in which the tour of duty is undertaken will be brought to an unscheduled stop only by a railway signal. Between stations her pottery water-jar takes on the nature of a sturdy little figure; a sweating, bulbous, sole companion. Moulded on its sides, two lions' heads, so Greek in style as to suggest truth in the myth of a journey this far by Alexander the Great.

The imminent approach of dawn rouses the traveller to the stillness of the train stopped in a new place, arrived at some time in the night. The carriage door, opened on the side away from the platform, allows in a wedge of early sunshine. Sun makes golden the matting on the floor of the square compartment. The touch of the golden light gives back to the woven grass a vital nature. The clean-smelling piece of matting lies flat on the floor plane which the occupant can judge is elevated above the tracks. The broad-gauge tracks are on the railway bund. Straight miles of bund formed by countless basket-loads of earth.

Although a traveller might be wearing only an embroidered cotton wrap, confident of privacy given by such isolation, an interested person can choose to be an early observer of acres of India in the first light of morning. On this journey, this carriage door faces the fast strengthening sun at the bloom of the morning. Later, the door, diagonally opposed to the first, opened, allows air to pass. On this, the platform side, no one but the tea-boy.

Soon, sounds of the breakfast servants stirring, calling out to one another. Not time, yet awhile, to set the table planks on trestles on the platform, to lay them, and unfold the chairs; or to start the station punkas moving. Time to signal up the tea-boy and take early morning tea, surveying the soil of India; to stand there, cup and saucer in hand, door open on a fresh place, with the promise of a new day. The possibility of a chance meeting to look forward to. Meanwhile, hot, sweet tea, and, laid out before the open door, the vast expanse of satin'd dust, the colour of powdered cocoa in the delicate morning sunlight.

The distant view on such mornings, misted over by the exhaled heat of the night, looks like the steel plate of an etching still in the acid bath; not quite detailed yet; surroundings separated from the viewer as if by an invisible, working – almost corrosive – liquid; the longer looked at, a reverse of the expected: the more light, the less detail.

Occasional snuffles from the engine; now suprasternly silent; next moment sissing; in steam, but as yet just dozing. Engine boys work, wiping over the casing and metal pipes, crouching on the curved, sun-shined black plates.

The compartment begins to heat up. Descended from its close confined heat, this one traveller now waits in a narrow strip of shade blessedly cast by the train.

To see out of the train's deep shade into the shade of the platform's canopy the passenger has to peer through a bright slanting plane of sunlight. To eliminate this sense of poor position, due to poor vision, the waiting passenger steps forward, passing through the plane of sunlight cutting down between the carriage roofs and the nicely judged position of the station canopy. The sun's heat is felt in passing. On thus acquiring a normal range of vision, the passenger is face to face with another passenger. He speaks.

"You were able to find a place on the train as you hoped? Persuade two men to double up?"

First a total stranger on the verandah of the club… then the suddenness of this encounter; compounded by a strange sense that this incident, in relation to a previous sighting, has taken place before… yet no true memory remains of an outcome; nor how such a precursor of this event could have ever been.

Some background music – like the twanging of India-rubber bands – permeates the memory irritably… so often the heat the cause of mis-persuasion.

The train stands lithic in the heat. Observed from within a compartment, thin bars of the landscape show between the blackened green of the carriage-work's louvres. Over all is a pall of silence, persecuted somewhere near at hand by flies.

Out in the sun's baking, nothing to speak of will move until the afternoon light begins to wane. Then worthwhile noise will gradually increase about the train and on the platform side will start up all the scraping and doing sounds of local life.

As soon as it can be seen in its true colour, freed from the flattening light of high afternoon, the ground looks as if it had been walked over everywhere: traversed by innumerable feet; pack animals or wheels; families or

beasts; individuals or fast-trotting dogs in the grey of dawn. The dust is criss-crossed with tracks, each lapping the other; lacing without any discernible sense of pattern. The dust will wait, a mish-mash of records of passings, to be again and again padded on, until – after the weather has broken – a new beginning in the mud.

As it is, the dust shimmers by day in a faint steaming haze; stretching away under the widely – wildly – dotted trees. Ten days of the tour of duty have shown her a similarity of places which add together to so large an expanse of soil-dust that the sum is a vast impression of native journeys. What for?... Where to?... Yesterday, today, sometime last week, a month ago? Time makes no difference on this continent. There is no wind, has been none; there will be no rain for some time.

On the journey... either afternoon rest-time, with the train stationary... or perhaps as in the early evening, the few carriages on the move, ambling along, parting the heat that is rising exhausted from the hardening-to-cracking earth... such long travelling spells as these spent reading; or reclining, staring out.

The dotting about of trees must be a result of survival; so unlike any pattern required by a true agricultural cycle. This disorder of landscape is difficult to read due to the absence of obvious man-made divisions of ownership. No visible connection of pieces of land to inhabited place by a network either of built walls or hedges and ditches to which trees could be bystanders. This immense landscape totally devoid of paths guided by up-heave or significant break of surface.

The traveller experiences a twinge of loss at the thought of paths which ran by walls at home... homeland, thrice marked, by wall, wear, and greener verge; in traces that ran up folds by easy inclines to brave the tops. Memories of paths once walked glaze her eyes.

Did the people living hereabouts know the bare soil of this blank land as those at home know the secret whereabouts of nests of primroses; dells cupping wood anemonies from the seasonal winds; sheltered berrying patches?

Instead of an easily read, varying landscape where all items are in their place and distinctive on the line of walk, here an individual could apparently forever wander every-which-way. Those who live in the dun-coloured villages may pass a familiar object differently each day for lack of substantial guidance. The watcher passing on the train has no means of telling. Even if the local people receive their sense of place from qualities of the dust, and far horizons, a passer-by would still have no necessity, because of the great space – no requirement at all seemingly – to be precise as on a path.

A path! True paths can only be remembered – that awful sea voyage cannot be contemplated.

A compartment built of wood, so recently that the enclosed air smells of wood; smells of the peppery cinnamon of the roughly tanned leather of the door straps, the studded upholstery of the divans. A compartment meant for four persons when hitched to an everyday train: two upper sleeping places could be made by letting hinge down two pallets whose under surfaces, visible in daytime, are varnished boards to match the ceiling. As upper sleeping pallets, glimpsed in the other compartments, these hang restricted on stiff, cochineal-coloured leather straps.

Directly off this room on wheels, a cubby-hole for washing in, containing arrangements more modern than any previously experienced. The ship's pinched cabin had been little enjoyed.

A bowl, with white-painted inner surface, projects from the rectangular tank that pretends to be a projecting panel of the partition; painted to simulate wood graining, with pretty borders lined in ochre, it is formed, and seamed, in the manner of fabrication of a tin trunk.

At every scheduled stopping of the train, up through the high central ventilator is to be seen an inch or two of thin, dry-skinned, ankle; a passing end of white cloth. The bearers of water-skins walk the length of the train on the carriage roofs.

Down the funnel of the other bowl, a frightening view of track can be glimpsed, rushing underneath. Equally disturbing to equilibrium, the brass flower of the grating in the floor allows in wheel noise and a rush of air against an exposed ankle: best to avoid stepping near this altogether. The woodwork of the windows is perhaps the most ingenious of all the devices the compartment offers the new occupant for the control of her comfort, being entirely suited to the climate. What kind of man in the railway yard had thought so well to consider these?... fly-screens alone? – fly-screens and louvres? – or solid panels? – never louvre and solid for these are in the same groove. All screens pushed aside allow the occupier to lean out, on the side of the train free of smoke-puffs – but grit stings the sweat-softened skin – she closes the solid panels, applying an extra push to close tightly to keep the dust out.

Soon it will be dark. Some time before, the lamp-lighter had passed down the swaying carriage tops, and lifting up the metal dome of the ceiling lamp high over her head, had, from above, lit the mantle, removing his light from its jar to do so; lighting up for only a moment a view of his being there before the dome hinged down and its white-enamelled surface reflected brightly all the newly-lit light. The presence of the light gives her an impatience that she should be better employed. Light too high up to read by. Uneasily her regard strays from the pages to the corners of the space; even to behind her shoulders: vainly trying to fill time until she undresses, brushes her hair a hundred strokes.

A pull on its chain puts the light out irrevocably. The night-light in its glass jar burns meaningfully for the first time since dawn. A weak light yet glinting in the metal disc that is the small protector of the vertical boards that warmly smell of the varnish applied in the railway yards.

The train, during some hours of travelling, passes only occasional villages, but as the night outside has become black-dark, none can be seen by those on the trains. The rail sound should have reiterated get-to-sleep, get-to-sleep. The train rattles and creaks, and even perhaps hurries on its busy wooden sway. Once the engine whistles: is there another train?… one passenger's ears are strained for a sound of an approaching train… surely the line is a single track? After a while, the train noise changes; the sound of its movement contains slurs… is the train slowing? The girl, kneeling now, moves the shutters to fly-screen and louvres; watches for signs of buildings, anything, on the line side. But nothing is to be seen. No longer are any lights on in the adjoining compartments to throw even a slatted, wheel-side band of light on any earth passed.

Irresolute, she stands on the grass matting they had ordered to be put down on the floor… the train movement is now interrupted by retching motions – displeasing to be reminded of the ship – definitely stopping.

Steam hisses, but otherwise no sound. The absence of sounds cause her to tip-toe about on the matting. Kneeling up on the divan to look down through the louvres she perceives dimly lit slivers of a platform's crusty surface.

The compartment door is allowed to swing open on its leather straps; carried by its own weight. The station thus made visible is a large one with punkas hanging among many cast-iron columns; wooden buildings varnished or painted brown. Beyond these, a farther platform. To the rear of the train, no carriage casts evidence of a light, nor projects an open door. The passenger descends, holding the surplus of her skirts in one hand, grasping in her other the vertical brass rail.

The engine is being tended under flares.

Immediately forward of the carriage ending in her compartment, is one ochre-coloured. The leading two coaches, first-class and white, are the official day coach and the solid-sided official sleeper. By a palely lit step, stands a green-sashed jemadar.

Nearer to the girl, in a position under a greasy station lamp, stands an evening-suited passenger who smokes his cigar.

One hand behind his back, this passenger watches an advance towards him from the only open door in either of the accompanying green carriages. The man wonders what the sword-bearing jemadar will do on this occasion… remain aloof?… or send an under-servant for instruction?

Miles from where the watcher in the train had experienced the arrival of dawn, the train stands in another station with a view of an almost identical landscape. When the shadows grow longer, the milliard crests of the powdered dust – having been flattened all day by the glare of the light – now appear to rise, almost as watched, and each casts its miniscule splotch of grey shadow, accentuating further those forward particles which catch the colour of the sun's setting light: each shining particle a-shimmer as the heat of the earth gives out to the zenith.

Thus arises an air movement of sorts.

In the sunrise – after the train will have departed – the light will shine from the other direction… and so on… this way and that way, for endless days. In the mornings, silver shining particles on a mist-grey ground instead of shadow-sharpened evening granules the colour of Huntley's ginger biscuits. This evening, the breath of exhaled heat is rising quite swiftly from the earth, as if trying to imprint its passing strength on the observer's face… is the place begging not to be forgotten amidst so extensive an area of similarity?

A land so dried up recalls the pomegranate story.

When she has walked sufficiently forward so that the station building's nearest sign – *First Class Ladies Waiting Room* – conceals the impressively uniformed jemadar, she pauses and asks the smoker of the cigar: "Why have we stopped?"

"We always make a stop at night."

"So we may sleep better?"

"Possibly." He adds, "Principally to adjust our timing between the last and first stops of the day."

The long-skirted passenger begins strolling on the platform. Slowly she circuits the man. The man watches; to do so he is obliged to rotate, turning on his heel to continue facing her.

"Do we stop long?"… "The train will not steam off without us?"… "You are sure?"… "You have never had to run after a train?"

Answers are formulated in pairs of words; or single assent; or denial.

"If you do not wish to be disturbed, I will return to my compartment."

The man shrugs his shoulders; the platform is on normal occasions a public one.

The man takes a look at the cigar. Next, despite the encumbrance of the hand by the cigar, the bridge of the nose is pinched between third finger and thumb.

When he has done with that gesture, he sees the girl is still there. She balances, drawing in the dust of the platform now with the toe of her right slipper, hands clasped behind her back.

"India smells nice at night. I like the smells here."

"That is my cigar."

"Not during the day."… she looks away.

He stares over her bent head, down the track beyond. Hardly a sound: the crew have probably finished their work on the engine and are not musicians like the crew on the last tour of duty.

The passenger wavers in front of him; willowing about as if on weak ankles. Initial worries at this unsteadiness prove needless; she does not fall down.

Nearby to where they stand are the two long green carriage sides; no sound comes from any compartments of the train.

The moon gives a bluish bloom to the high central roofs running the length of all carriages. The clerestory of louvred ventilators casts a heavy shade on the curved surface of each lower roof. The angle at which brightness falls from the moon causes the heavy shadow of the hood mouldings of sunshades to appear to be eyebrows over the relatively insignificant train windows, and there to accentuate the forehead aspect of the blank band of above-window carriage-work: low-slung ventilators between the wheels carry the grimace further, making the coach-long mask extravagantly bewhiskered like a kathakali dancer's painted face.

Two separate ochre-coloured carriages, one for the servants from the Residency, the other for the Hindu breakfast servants; its solid-sided van the store for food, cutlery, cooking apparatus, camp chairs, table trestles. Although each larger station has its regulation menu to be served whenever would-be patrons telegraph ahead, the personages on this train have servants with the means to take care of all their needs, whether the stop is made in small or large stations, or in passing-sidings. In the two green carriages sleep individuals whose interests conveniently conjoin the yearly circuit, and who otherwise would not have the benefit of the hindmost horse-carriage.

The man standing on the platform takes all these occupants into consideration before he turns resolutely about and walks away from the unsavoury choice of who might overhear any conversation tonight. Secretaries, aides, would know his voice and be curious; he could not judge the reaction of the auditor to the Bank, the two belonging to the Company, the newspaper scout; nor those two Inspectors of Education or Salt Dues.

The girl does descend from her compartment, and walks up and down the platform. "When peacocks land, their tail-sway practically see-saws them off their feet… particularly when they fly up to land on the top of a very high wall, you think they are about to fall off. Their tail shows most splendidly draped over the surface of a whitened wall."

The man has no reply, but she can make out his whereabouts by the unbleached linen suit, although the proximity of the station buildings makes shadowless the fabric's folds.

"How long have you been here?"

"In India… or here?"

"Not on the platform."

Not used to being laughed at, the man turns and retraces steps taken only a few moments before.

The train rarely maintains any speed; idling along. The air, blowing hot, passes endlessly over the carriage-work, carrying on its breath incredible quantities of grit and smuts. Air also laden with such fine, fine dust that even the compartment interior, closed against the dust, is fogged, the sunbeams moted. Yet all the windows, as the doors, are tight shut, bolted against the threat of robbers known to board moving trains to rifle baggage. On some days, with each joggle, dust puffs, as smoke, through the working chinks of carriagework.

This morning, the louvred shutter raised exposes to view a rocky, towering landscape. Commanded everywhere by forts and monuments whose presence – everywhere she stares – compounds the shock experienced at the dramatic transformation of scene. The endless plain had numbed expectancy of change.

Sun-bleached, distant tops: baked stones watching the train move; surely here too slowly.

An observant riven terrain, each moment more leached of colour by the increasing light of day. Gradually any indistinction of topography gives way before all-over sharp detail. Objects so clearly defined as to encourage urgent search for human, native observers.

The residual warmth of the compartment's wooden surfaces has impounded air that has been endured all night. Now this is budged. The emissions of all surfaces are wafted back a breath as rock-stored heat advances across the valley floor and, as if to impress its vulnerability, passes through the louvres of the train that is proceeding at so leisurely a pace: wooden sides no barrier; insufficient movement for escape. Over a period of only minutes, the watching passenger sees grey washed away from the scene, true colours shown. Even to the picking-out in colour of sparse leaves on aged trees; detail in turn retracted by the sunlight.

Then finally – as on each preceding day – mere residue tones, although here worried by rock formation, for the rest of the ride of the sun.

One hand, holding the jacket at the turn of the hem, holds the front sufficiently open for the right hand to take out a pocket watch… the man's head is held in such a position that the eyes seem to look down the length of the nose at the watch face… as with so many long-service writers, the man's face is of a tone with the natural, native weave stuff of his suit. The man's presence induces in her a sensation of being out of her depth. Please! – the thought pleads in her head, as she stares at the stripes on the shirt above the waistcoat – please!

The time, after midnight: the thermometer on the station platform – despite the sound of a sharp tap with a finger-nail on its glass – stands at 109 degrees Fahrenheit.

"Some people do not seem to enjoy being here… do not seem to care for Indians… the climate or the place… do you wonder why they do not go home?"

"Just a pose. What else are they fit for?" Arm outstretched, a forefinger taps the cigar which discharges a plug of ash on to the grit between the rails: grit enamelled grey by the moonlight. The man, turning slightly, looks up the track, then down, either way, the metal rails shine until a thumb-width apart, then continue unreflectant, discernible only on the lighter grit of the dark form of the bund heading straight into the shades of the horizon.

Much later, in a considered gesture, the butt of the cigar is thrown away. "You should get some sleep now", is said kindly.

A whole group of beautiful domed pavilions; set out in nowhere. The compound wall so broken down that as the train slowly passes the passenger can see the composition through and through and out the other sides. So many fine buildings, not part of any establishment or community, nor near a trace of ruined occupation.

The sun reflected in the tank takes on a form which is sometimes gold-embroidered on Maharajahal pennons; or beat-out of brass for processional elephants' breast-plates; or carried as trophies on tall wands: a small tight face surrounded by wavy lines radiating to three or four times the diameter the grin might accomplish.

Perhaps it is some sound that causes him to turn.

She stands not five paces away, yet is not regarding him.

"Oh, we are outside a town tonight." The ordinary spoken of in tones more appropriate to a change of scene at a pantomime.

A village composed of cumberous, primitive, grey cubes; perfectly innocuous from one end to the other.

"What is it called?"… but already, the station name board is found, quadrisyllabic inscription read aloud.

He corrects the pronunciation without much thought for her feelings.

"How often have you been home?"

"Once." This said looking out across the village; the flat roofs, close below, are almost within a leap of the platform's edge. "Went back after six years; not since."

She waits. He says no more. "Did you lose anyone in the Mutiny?"

"Not really: a few perhaps who joined the Delhi Force…"

The idle question referred to the time he had been fresh out from home. The man then had not known anyone well enough to now recall individuals clearly; his memory presented faces without names; names without identities. A muddle made the more distressing by the inconsequentiality of her questioning.

She notices his hand is pinching the bridge of his nose between finger and thumb.

The moon shines from so high up and so brightly that the sky cannot be seen, unless it is imagined as a far-distant blue-grey hemisphere, upturned over the great open plain: a dome of night situated well behind the moon.

Both strollers, their heads turned to one side, seem intent on staring out over the village. The platform edge dictates the straightness of their line of walk and gives them vantage over the moonlit housetops and unearthly soil. The surface they walk on appears to lead out like a great pirate's plank over a delicately shaded yet moonlit void; their position apparently not quite of the earth; although, to the feet, too solid to be as ethereal as its semblance.

I shall have to go – she was thinking – I cannot stay out here indefinitely taking the air or he will know I do so to speak to him, not for sensible, unconfined pleasure, or for my health.

A slow walk; a turn; and back again, very slowly.

Turn again… for her, this walking together is an experience requiring concentration. She hopes to appear at ease: if she is fortunate, there would be eating with him, driving in a carriage; all sorts of activities, and this the person with whom she should feel most at ease.

She lags behind to run her toe over the platform surface where, repeatedly in their walk, a difference has been touched on.

Noticing her pause, he glances back… has she seen something disturbing?… He looks around for any sign of movement although ashamed of his suspicions, for he knows thuggees only attack sleeping travellers whose party they have infiltrated over days… A pack of dogs perhaps?

"What is this? There was one at the station where I first spoke to you."

Thus are a casual few seconds, emptying out the end of day of routine, made to sound like a landmark on a journey.

"I can feel a surface mark, with my finger." Crouching down, skirt bouffant. Hitching up a trouser leg… The man flicks the burning fuzee out. Momentarily, in apparent darkness, it seems they might bang heads on their both rising so suddenly. A hand had been offered in assistance but she had not seen it. He walks away a short distance before pausing… strange, women remembered such incidents as first times, last times, quite small events, meaningless in themselves… meditatively a foot is placed over the ash fallen from his cigar. The foot removed reveals a light grey smudge on the platform's surface.

Two passengers separately stand looking out over the village. Rag-covered stalls at the bund foot – somewhere an animal noise – voices, suddenly shouting can be located in the middle distance – no lights. Then silence; seeming greater than before.

During the day a great hill fort is sighted by the passenger, alone in her compartment. The chatris on the fort's roof are outlined against a sky too hot to be described as blue. Rivulets of curtain-walls, run down and up and over the hills. As the train passes, the walls show momentarily as crest fringes, reminding this observer of the way a sheepdog's hair parts when a cross-wind blows. The comparison, so natural from the past, now made unnatural by circumstances, causes a twinge of regret. She moves restlessly, to shake off a lingering sadness.

Today's villages are composed of mud, yet seem imposed unnaturally on the ground; as if grey slabs of peaty texture have been set down without reason; as if their siting on the grey earth, further off, ten or twenty yards in any direction, would have been of no consequence. The other passengers would have seen all this before… must think her mad to talk about what she sees from the train.

There is no personal servant to perform tasks for this isolated passenger. But no great pride is thereby lost; the few clothes and effects can be arranged without help.

Untouchables come to brush the floor, to clean the bowls; climbing in and out from the line side. Entering while those in the non-official carriages sit out at breakfast.

The passengers who sit in the open, disposed either side of a trestle table set in the centre of the platform, are serious breakfasters.

From inside the carriage sounds only a whisper of sweeping; the bristles of a whisk show momentarily in the cleaning of the louvres. One sweeper sprinkles water on the floor, finger over the neck of an old glass bottle. Another brushes only, wielding a lop-sided bunch of brittle grasses so worn as to need three times the gestures a freshly gathered bunch would make to clean. In turn follow the other sweepers, each to carry out his own bitty task with pan, stick, brass cleaner or oil.

Steadily pounding along, mile after mile through the powdered ginger landscape that stretches to a watered-silk horizon.

The girl lies longing for, then hangs on to, the sight of any slight undulation. However, her peace of mind is attuned to the beguiling stillness beyond the travelling sound. To be so many thousands of miles from that perfected quiet of the moors above home never lost poignancy, despite repeated remembrance.

The thickening luminescence in the atmosphere at the approach of dusk a good exchange for the sea-haws of home.

Hour after hour the train chuffles along, its noise feathering towards dawn.

Something akin to this feathering, but happier, sounds if the engine pulls in the early evening. Occasionally at this time come cross-breezes and eddyings in the rising heat, to snatch away the sound in happy gulps.

The too-ripe apricot gilding of the tree faces catching the light of dawn contains a foreboding of the full heat of day. Dawn colour touches objects momentarily.

The light of the sunset placates. Even the air's touch is softened by the dusting of earth therein suspended. All the final tints of sunset are appreciable. The brightness boldly resisting fading burnishes all objects with the bronze of the light's battle to remain.

Often the landscape which the isolated

passenger looks out on epitomises tiredness; the very leaves look worn out after the ravages of the day's heat.

By the water-tower the engine is being tended, but the flare of light from the work torches does not reach anything like as far as the green carriages: the man notices this fall-short of the flares as he walks to a position obscured from his fiercely moustached jemadar.

Now, his back deliberately turned on the train, the man smokes the last cigar of his tour of duty. Again a platform is high on a railway-works embankment. Again, below it a strand before a village. An entire village – walls, ways, roofs, rubbish – all formed of the same dirt. All day the train had passed villages tone-decorated with dung-pancakes. Slapped there by the women from a squatting position. Hand-cast patties thrown onto hand-smoothed wall. A wall-held store, at low level, within reach, drying to become fuel. An aroma of dung smoke mingles with that from his cigar.

The station is even larger tonight. The size of some of these stations presents a puzzle; what traffic determines such width and length and multiplicity of platforms.

The jemadar outside the white carriage is companioned by a chuprassi. Their figures are positioned formally, either side of the carpeted steps. But no light falls on the two uniforms, the carriage door is closed.

The train rests, shuttered against intruders.

Over the tracks, a double platform swarms with people. The oil lamps over the crowd are widely spaced. Their cages of interwoven wire are all but covered with moths. The light emitted is haloed by flying insects.

Whole families are encamped amid bed bundles and hand-made boxes. Seen in the open, under the swaying of the lamps in accordance with the motion of the punkahs, the moving obscuration of the light sources, the light's mottling of the shadows, the crowd appears to be seething, their movements threatening.

On the platform in the siding, two passengers take the air. The two are enshrouded in darkness around the railway-works and are not exposed to view. Nevertheless the man suggests, "We had better move down the platform."

In their moving away from the watchful railway officials, the squatting, eating, engine crew, the jemadar and his companion, she keeps in step. His stride, longer than hers, leaves her a shoulder behind after five steps taken. However, her own normal walk resorted to, bobs out of time embarrassingly.

How frightful if our train now left."...

An amusing series of images enters his mind, presupposing the attitudes of certain people as such a scandal passes around cantonment and Club...

"You have a nice laugh."

The shelter of some station premises gained, the man leans back against the wooden boarding. Soon, he hears her nail picking at the varnish that has blistered as if into pin-heads.

Looking up at the sky, he sees above his head, a stretch of pale stars, none of which he can identify.

The girl stands, no longer picking at the wooden boards; her gaze is intent on the gaudy crowd.

Her talk begins to enthuse about their domestic habits, their clothes; all she can make out at such a distance. Drawing his attention to those who stand up exposed among the packed throng, on and on... a stream of girlish reflections in a nervous staccato delivery, as if the observations are being thrown at the man.

"I am not boring you?" she asks at one point. "You will say so."

At a certain later moment, the man, in a movement beginning with a twist of his body on to one shoulder, heaves himself upright from leaning on the pimpled surface of the building. Pointing with the hand holding a cigar, pointing across in front of her, "I think there is a train coming."

As yet none other than the working sound of the waiting crowd.

"Maybe we should move back."

The stroll round the end of the station building is not such an ordeal.

Without much travail, her compartment is arrived at: the initial strangeness of walking together cannot be repeated.

In the new situation, the residual heat that all surfaces emit, the closeness of particular surfaces, create a sense of their being confined between the sleepers in their second-class green carriage compartments and the station buildings. Whether a sleeper, awakened by the noise of the train arriving, will be able to see that part of the platform on which they stand, will depend on whether the awakened sleeper is jolted into kneeling up; the lower louvres would obscure a view from a position raised on one elbow.

The girl looks down to estimate how visible are her skirts, or his suit. The white cuffs of his shirt indicate the whereabouts of his hands.

The train comes snorting. The oncoming tremors quickly become passing thumps. Rolling heavily past, blowing steam, a big brute of an engine. On driving wheels as high as she stands, the pounding bar on them, with

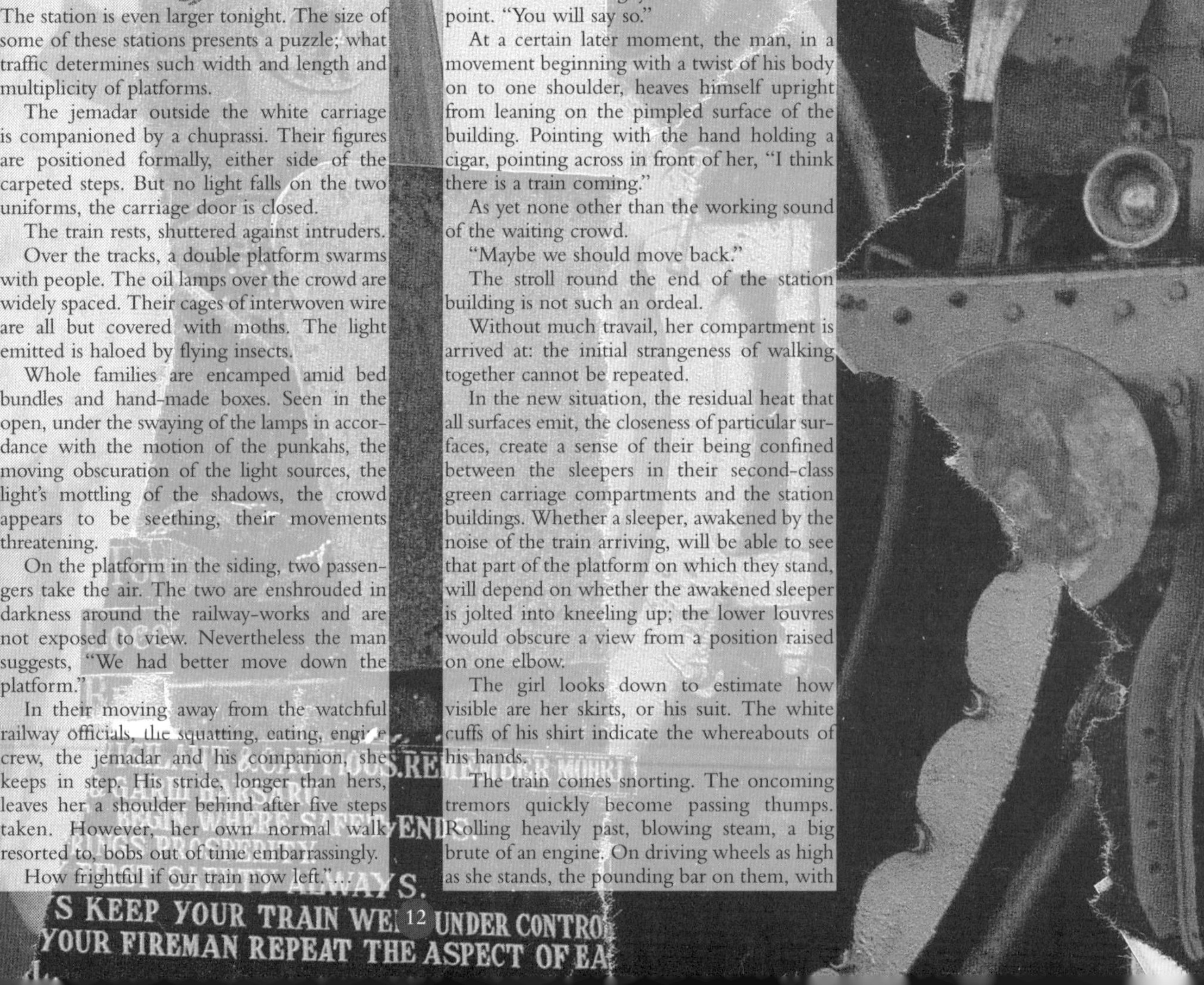

increasing effort, reaches as high as her eye. A cotton blind is flapping brown from the driver's cab.

She touches her companion's arm, suddenly alarmed at what she sees – "Nothing has happened, has it?" – arms dangling, heads lolling, travellers' bodies flopping out of window openings… The coachwork's ochre appears sickly in the night.

The hidden jostle soon stirs those on the side of the train visible to them.

Boxes and bundles are heaved about by unseen hands. The recognisable silhouettes of pottery water-jars are handed in by outstretched arms; water glints as it is poured over hands onto the track. Sound, at first muffled, comes more directly from the train as sleepers awake, fan themselves and each other.

She whispers to him, "I adore the people of this country. I never realised I would so enjoy living among them."

The listener remains a while longer, perhaps to be sure there is no more to listen to.

The evening declared over, the compartment door is closed behind her before she can reach round to do so herself.

Although underneath the tablecloth stand trestles and a plank table top, the bowl of fruit brought to their breakfast table comes ornamented with nosegays of roses, mogees, jasmine; gathered somewhere in the pre-dawn cool. Each time she rises from the table she takes up a posy so that she may sniff away the morning's dust.

At the evening meal, Japan roses, oleander, ornament the gindy and ewer carried round to each dining passenger. One night, in each perforation a tuber-rose; their dewiness completely concealing the water which had been poured over preceding hands.

Framing their last meal taken together, at a table corner, swags of freshly gathered leaves, fastened together with tiny twigs, as if fabricated by the tailor bird.

The girl looks round at the other passengers, chortling, eating, drinking the inevitable claret soberly enough. Yet the leaves fashioned as swags brought to her mind an illustration. It mattered little whether the servants bending discreetly between diners at table were slaves or natives. Set up on the platform, amid cast-iron columns of attenuated classical style, was a dinner table for men who were blind to the charm with which such an appropriate decoration had been interpreted.

The carriage door is drawn-to from the inside; he hears the anti-burglar bolt shot.

The man walks back to his carriage alongside the stationary train. He notices that drips from the cooling system are as a fringe to the bottom of the blank white side of the official sleeping car. The jemadar at the steps in his green uniform stiffens to attention at his approach. The man enters, hears the servant close up behind. The chuprassi will be left on duty outside; the choukedar sitting in the cubbyhole inside. The man retraces his steps, this time inside the train, passing the length of the day coach, through into the corridor to the sleeping compartments. His kitmutgar waits at the door. The light from the compartment falls on the white of the servant's uniform; the now bowed turban of the customary greeting.

The official passenger only comes to a stop when his knuckles rest on the dressing table top. He supports his weight on knuckles that press heavily on the glass sheet. He looks down on a conjoined pair of ivory brushes, a silver comb laid parallel to the front edge, its teeth towards the mirror. One hand reaches out and moves a tortoiseshell stud box on to the other axis of its oval. Then the man shrugs, as if to express… no matter.

Seated in the armchair to undo his laces – he does not ask his servants to undress him – first the boots are picked up and handed out to the under-servant on their way for cleaning.

Silk socks, then shirt, are tossed into the fresh cotton lining of the laundry basket which will be carried out last. To this servant he speaks Punjabi: servants only speak Hindi in common; Hindi is used for all household communication at the Residency.

To wash: he watches the engine-warm water from the brass faucet run over hands held beneath the outlet. Stares at his reflection in the mirror, turning his head this way, that way, and turns the faucet off: must get his nails cut again.

During the night, the man suddenly wakes. The night-light flutters on the partition. At first he hears only his own heart thumping as if at some awakening shock. After a moment's thought, puzzling what could have woken him, he turns over on his side and stares into the compartment whose gloom appears to be held a hand-span away by the night-light in its blue venetian glass chimney. Listening, he can make out a peon, or the chuprassi, snoring. The snorer will be out in the passage. The train creaks, it is moving.

Chapter 2
BREATH OF INDIA

The middle of June sees men on those roofs visible from their verandah. The crouching workers turn the tiles, relaying and adjusting them with flamboyant gestures. Thin arms upwards reaching as if the expressive hands would milk the sky.

The tiles clatter happily at the start of the morning. Rhythmically sounding, as if a clapped accompaniment to remarks called from roof to roof.

The sun climbs and the tiles chatter more insistently as if the heat emits its brittle voice. Or is the rattling of the tiles a rite performed with the intention of shattering the porcelain quality of the sky?

A few days, and all the roof activity is over. The pall of heat now broods, silently shimmering above the tiles. The brightness of the wavering light bleaches any trace of pink out of the roofs.

Visiting the dispensary, old India hands become jittery; sharp tongued against servants' movements, and the listless acceptance of fate.

Each day, buildings soon turn into ovens that over-bake compatriots' characters to friable facsimile. The girl feels very much the same person she has always been; yet, when the air's breath is hot, she contrives to walk more airily. To proceed so far as across a compound, to reach the gharri, she has to do so in quite a new way… shoulders differently slung to create space in the armpits; arms offset to hold elbows away from the body; gait more free, to move air under her skirt. In these conditions the linings of clothes stain so distastefully with the dust.

Each day, from its start, seems to stay on a well-worn course. As if the spirit of heat declared that whatever happened – gestures, conversation, toilet – would be blowzy. You got up, and before the heat of recumbence could be dispelled, the heat of the day beginning, clutched hold. Then, as if the heat weighed down the body, your person feels less agile.

Air heated outside creeps into sealed bungalows. The closed rooms fill to bursting with oppressive air. Beams of hot sunlight almost daily find new chinks to penetrate.

At unpredictable intervals during the day spent at the dispensary, wafts of heated air suddenly, swiftly, pass. On looking to whence these ethereal bypassers come, a little swirl of dust suggests that some baking of the sun's has erupted. Such a local, irrational movement of air, the body gamely tries to persuade itself, cools the skin.

A person cannot accomplish simple tasks that might have been done well in other circumstances. Had she not been here? She ponders on this. The visitor to the dispensary would never have been known to exist had she remained at home. Providential that she is here, not there, where this would never have been… most of the day it is much too difficult to think.

As the heat wanes, faculties become somewhat restored. By which time the day is almost over. Another day's time, she worries, is almost up, and what is to show for it? Tears moisten eyes whose lid linings sting more fiercely for an instant. Each day burnt up means yet another day has passed. A lifespan has one less day ahead. This sense of the wastage of the days crawls up the marrow of the bones of her arms, causing her to stroke her hands slowly up and down the tight sleeves of her dress.

Various events run by military stations of this size stay his departure; although the rooms in which the officers sit soon become prisons to their flesh. A body encased in uniform finds it impossible to sweat due to the over-wetting of the mats: even the face, the hands, remain dry; feel

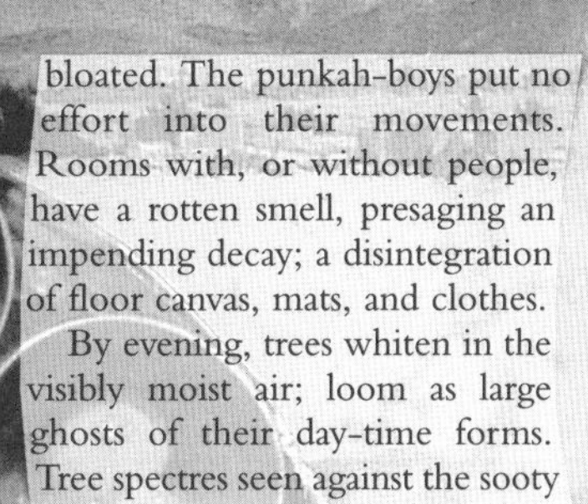

bloated. The punkah-boys put no effort into their movements. Rooms with, or without people, have a rotten smell, presaging an impending decay; a disintegration of floor canvas, mats, and clothes.

By evening, trees whiten in the visibly moist air; loom as large ghosts of their day-time forms. Tree spectres seen against the sooty backdrop of the chirping night. The humidity hazes into wafting curtains, as tangible as muslin, hanging among the isolated trees on the lawn. The verandah boards outside the doctor's spare room are a velvet of shining droplets.

The same overpowering humidity that now pastes his clothes to him has worked many years at indelibly mildewing the texture of the leather – like a torn-open orange – that covers the bedside Bible.

For weeks past, the local people, in their houses beyond the high mud wall bounding the cantonment, have made the unmoving heat of each night hideous with their fast, repetitive callings to their beating of drums and kettles.

Each day, more unfriendly look the white-washed faces of those compatriots who never move outside without bearer with umbrella, or cream-fringed canopy to their carriage. The punkahs swing, from nine in the morning until six in the evening.

This is the pattern of her first three months of residence, while the temperature climbs steadily. By the end of the first half of summer, water blisters on the palms of hands; and wherever clothing can possibly chafe, it does.

On the hot side of the dispensary, all the doors are fastened back flat, replaced by tatties hung on rings set in the wall above the lintels. Old India hands visiting the dispensary verandah, claim that tatties, kept at just the right moisture content, make 18 degrees difference in temperature: most tea times this opinion is passed; the repeated tellings of the information as if she had not materialised the day before; the day before that.

At the start of each visit, these men first stand before the white-cased barometer and tap the glass with their finger-nails. One visitor raises his gold-rimmed spectacles to his forehead, the while thrusting forward his chin .

Her high desk she has instructed the bearers to move to the only seeming path for air between the entrance steps and the continually opened and shut door of the doctor's room. Under the desk lid, the tall, narrow-lined books – one for each dispensary day – have many of their blue-ruled lines watered by her moist hands; used pages blown apart. The unused pages lie unblemished and tight-packed. The three sides of the firm block of unused pages are marbled prettily in sea-green and maroon.

Endlessly wandering round the verandahs, the tattie-wallah, in splashing water from his bowl, makes little slurring noises of tin drawn across the bowl. The little tin has been rescued from some kitchen rubbish; still embellishing its side are the emblems of a home brand. The noise grows fainter slowly; until unheard. Barely audible scrapings as the wallah returns to the far verandah corner. The approaching, regularly repeated action, increases in sound to clearly heard slurs. Then gradually again, slurs of tin passing across biscuit-ware bowl, grow fainter. Soothing sound fluctuation repeated until a person almost ceases to notice the figure's passing, or absence.

It is said that in the better houses the tatties are of aromatic grass, the fragrant cusa. It is said – at tea time – that in the bead and bobbin latticed merchants' houses, deep in the city, there are tatties that are sprinkled with attar of roses. The girl wonders if the tatties so treated are without the now familiar smell of mildew. Full-fluted notes of the bulbul continue unabated through the hour of sunset. Dusk is now approaching with an almost physical presence of a fully draped goddess bearing down upon the heat.

A first breath of false breeze. One of the rings from which the tatties are suspended gives out an encouraging frail creak. Day passes swiftly into night. A temperature of 121 degrees in the shade is allowed at last to sink to 115 Fahrenheit. By night the mattress is burning against the back. She lies in the darkness and feels as if she has shrunk to being smaller than life size; that she is wilting on the vastness of a continent.

It is important to her, each afternoon, that she tries to gather up sufficient energy for the early evening so that there will be a collected personality; some spirit with which to move prettily about. Yet each afternoon is too like the last: all afternoons tend to merge.

Even if behind closed shutters she moves about, and spends the time in pursuit of order, the girl is tormented by a maddening sense of the clock's minutes tiptoeing away with her youth. She drapes a shawl over the offending timepiece to deaden its insistence on the threat. She is fearful lest her life is thought to be of no purpose; that her time is seeping away with as little meaning as native time. She is distraught to find she is left with insufficient bodily force for even the most ordinary acts… her limbs now have no strength… her grasp has lost its power.

The cocoon of the linen hammock confines; leaves her without the will to stir; bereft of sufficient imagination to think how she might better recline. The long rope beyond her feet stretches out unshaded in the heat. The drinking glass and the sweating tan water-jar, both covered by a blue bead-weighted cotton gauze, stand on the tambour-framed garden table: within reach if she puts out a steadying toe upon the ground.

The birds never seem to tire. Their racket fills the tiny garden.

She rests very still. Does not attempt to read. She conserves what energy she has; and muses on the visitor. Might he again take his tea at the dispensary? Might his visit distinguish that afternoon?

Birds hop down, out of the trees, and back up again, with undiminished energy. Her stillness encourages the most timid to continue pursuing their affairs.

She rests, composing herself for better behaviour.

Plump: a common green pigeon lands. Pigeons look like pigeons at home although yellow and green replace the grey and mauve… there are also blue ones.

A number of species mutate, colours reversing or brightening, yet remain sufficiently like those birds at home for it to be disturbing for her to watch them here. Did sparrows, for example, come off the boats?

Other species quite change their character… blackbirds here are ashen… birds called thrushes not at all like those of home. Whereas a bird that could be taken for a faded thrush, here is called a babbler.

As to the shrikes; where has she seen their like before?

Some birds are gaudy feathered out of all belief: made half of yellow, half of blue.

Persimmons glow in the garden trees. So many orange balls amid dark, shiny leafage.

With each second that passes, the fading light increasingly magnifies the bright colours of the miniature exotic orchard… fruit… flowers… birds… each thing jumping into sharper tones as the light is muted.

Such a noise penetrates in from outside. The girl considers the air-borne sounds. It must represent so many birds. How can there be anything left over for the birds to eat?

The girl no longer concentrates

on reading the volume on the doctor's desk. She stares at the gauze-veiled vegetation of the courtyard. She gives a slight jump as the door to the room is opened.

After a nod of recognition and a moment's reflection, a comment is offered." Surely too big a book for you?" The man's knuckles rest on the desk's other edge.

Indignantly the girl replies.

"I do not mean anything in particular." The fingers of the visitor walk along the moulded edge. Finger tips slide slowly on the satin grain of the mahogany. The grain is polished. The shine the sum of years of elbow grease. In places the wood is stained with spots of oxgall ink. Other splashes are the fault of some past dispensing.

"This is the Sanitary Commissioner's Report; really Miss Nightingale's work."

"A person who was never here."

"Have you read this? It is powerful reading. So sensible."

"Do you understand it?" The man raises his head. He now rests his thigh against the edge of the desk. The man judges correctly. The four-square desk, whose pedestal cupboards are loaded to bursting, does not budge. The desk receives the full weight of this tall man as he leans. "To learn Hindi might be more useful to you."

"It will be useful here to know how Miss Nightingale thinks things should be."

The man averts his gaze from the tense expression on the face of the girl. He looks out through the window opening, at the sun reflecting on the dark foliage outside. The frame that holds the bellied gauze is set deep in the embrasure. This setting of the gauze in deep shade hardly veils the scene. Native plants and pomegranate trees weedily crowd the outdoor space.

The room the man stands in is small. So close are the walls that, left arm outstretched, the tips of the man's fingers now press gently on the bellying of the gauze. Innumerable expansions of the gauze have lost it tautness.

"I mistrust smatterings of anything... one has to be a fanatic to make even the slightest impression on this continent; the slightest difference."

The heat begins to build up with the coming and gaining of full light. Its first omnipresence little alleviated by the smell of the potted plants being given their water for the battle of the day.

But soon the smell of cooking – which surely nowhere else smells-out so clearly each tiny source of heat – counteracts the cool suggested by the smell of watered plants. The over-burdened air can do without these meagre additions of the charcoal's heat.

Hot season; to be followed by one humid, more enervating. Which season in turn would be mastered by increasing heat. The year turns on and on, relentlessly winding away her life.

Even in the shaded house – all openings covered, hangings moistened – some quality in the cloistered rooms arouses a nauseating sense of desperation and leaves the flesh in tattered rags of wanton willpower. An atmosphere too powerful to recover in, forbids rebound; presses on her body and squeezes out all resilience; leaves her no more strength than a fly... leaves a person no more effective... no more personable than an insect. All around her in the quivering heat, India is eating people up.

Only too willing to stroll apart during the evening promenade of vehicles... across the parade ground... across the light stream of horse traffic in the Mall... down the rotten ground that forms the river strand. The sky ahead is lit by an afterglow of Macedonian gold. Reflected light skims the pools of the dried-out river with an unhealthy metallic sheen. Bruised colours can be found everywhere. Except where they should be, in the sky.

The sky is far too hot in light.

Up on the Mall, the turgidly moving carriages, the promenading people, pass, and repass. The two strolling on the river's strand see those on the Mall in silhouette: the Mall's flares have come to prominence as the light has faded.

The two strolling figures seem sunken in the sharp grained dusk of the river strand. They proceed at a very genteel pace. Walking slightly apart, not talking. Treading the silt as if just about able to make their way in the dusk-thick air.

The noises of the Indian city sigh repeatedly. As if voicing a desire for a let-up in the oppression.

At the passing of the two people a few fowl cluck from their roosting by the dhobi men's huts. To the European nostrils the smell of curry hangs heavily about the scattered huts.

"Do they draw bright red hand-patterns by the doorways hereabouts?"

"Little twigs at the ends of fingers; birds coloured as robins with twiggy feet." In answering, the girl pauses, looks back towards the last dhobi man's hut as if to indicate what she has noticed. What happens then is not at all as she had imagined it would be. She has been nervous just walking beside him. Now, given the opportunity, she wants to please but cannot recall in time the emotions so often imagined in hope. In a moment, and in a fashion unprepared for. She stands stiff with inhibitions never suspected. In an awkward pose. Her mind is ridiculously, disgustingly, on the position of her feet. And what to do with her hands. Before anything else, the occasion is over. He touches her elbow, to indicate they should walk on; as if nothing need be said.

Mindlessly the girl gazes at the river's mirrors, seeming as solid mercury in the evening light. Areas which by day are buff-coloured pools of over-used water. The flimsy seasonal shelters of the dhobi men that are sited close to these shrunken pools stand also on the brink of a desperate situation.

The man studies her profile in the remaining light of the sky that faces them: she looks content enough.

The girl had neither felt sufficiently receptive nor encouraging: in her desire to encourage the man there was no idea of her being responsive, purely the right nuance of acceptance.

Circling kite hawks and heavier crows are sailing over the military hospital. Similar slow gyrations in the pewter sky mark out the locations of the barracks.

The doctor considers the military let their soldiery live in dirty conditions. Men eat from wooden bowls rinsed in rarely changed water butts at the end of each stucco-brick mess-bench in their yard. Any cooking is done on the ground outside their quarters. Bearers squat on their haunches to make up food bought in the bazaar for individuals or groups of friends. Perhaps worse, is the practice of the soldiery to stand eating food bought-in already cooked.

At last for the hospital – thanks to the Sanitary Commissioner – an issue of enamelled plates and mugs.

In the military hospital, an increase in deaths due to heat-stroke although punkahs hang within three feet of each bed sheet. Punkahs which began working 29th March should stop on the 29th October.

About the middle of July comes news of the monsoon arrived in the hills. There, she is told, for five to six wonderful weeks, the long-awaited rain will mist the visibility, reducing it to ten yards.

The girl, as she listens, declares that she would some day care to visit a hill station. They explain to her that a second-class hill station will be some 10 or 12 degrees cooler than these plains.

In mid-August comes the astonishing rain. Just falling in

CIVIL LINES

भारत के सब प्रदेशो की सफेद और रंगीन खादी ।
कोटिङ्ग शार्टिंग
विविध रंगों की छपी सूती और रेशमी साड़ियां ।
हर एक प्रकार की सूती और रेशमी छीं
धोती तौलिया चादर रुमाल
टेपेस्ट्री परदे टेबल क्लौथ
रजाई तकिया होल डोल ।
जाजम गलीचे कशमीरी गब्बा ।
बैलघानी का तेल (तिल्ली सरसों)
अखाद्य तेलों से बना हुआ साबुन ।
ताड़ का गुड़, चीनी और अन्य पदार्थ ।
हाथ कागज और स्टेशनरी ।

सात लाख देहातों की आबाद
प्रोत्साहन दें। हाथ से कते सूत की
ही आग्रह रक्खें।

straight rods. As if pushed or weighted. Met by the mud jumping up as if spurted out of the earth by an equalising force. Rain in sufficient haste to send a continuous silver sheet leaping over the gutters. Rainwater arching down from the roofs to waste and dirty.

The shiny surface of the floor glistens as does a sheet of water in horizontal light of evening. The sheen suggests quietude, recuperation. Thin white legs of terrace furniture epitomise airiness. Lamps, hanging among baskets of ferns, accentuate the bright green of the fronds. More curling, lacy ferns, in whitened basketware stands, uphold this image of a verdant fringe. Therefore, with a heightened expectancy of refreshment, and a considerable sense of relief, the visitor ducks through the protective layers. In doing so, he comes all too suddenly upon a girl who had been concealed behind a white wooden post. A list, clipped to a board, is balanced on her arm; the pose, ribboned cap over tightly-drawnback hair, is of a school marm. The pose strikes the man as slightly ridiculous as the girl looks barely old enough to be a pupil-teacher.

"You are too late for tea; when the doctor is free he still has all these people to see." The speaker fixes a preoccupied gaze upon him. Her free right arms swings out protectively to indicate the floor's occupants; families whose squatting poses show not the slightest sense of urgency.

He laughs at this sense of time fresh out from home, "You sound old-maidish."... Her response happens so quickly he is caught off-guard. He realises he is not as nimble as he thought. The presence of the hall of families holds him in check, expressionless.

No other consideration has time to enter his head before he hears a hearty "Look here." As the recognisable voice speaks out, and hurried feet approach, Nicholson moves adroitly. Together the two men go into the doctor's room and close the door.

Nicholson has no wish to consider the matter further.

"But in front of the locals."

"Dougal. The girl will suffer enough for it I don't doubt: all round town by this time tomorrow, if the tittle-tattle fed by servants is here anything like other stations."

Undecided when first invited, he judges differently on the night. Nicholson suspects he will rather relish springing a surprise on the cantonment's gossips. He has no stomach for letting the cantonment feed on the girl. He will treat her with the respect due to a protégée of Dougal's. On entering, he recognises her more easily than he thought possible. On his first sighting the numbers of young compatriots, he sees her looking very young in thin white stuff. An unfussy arrangement that takes her tartan sash Balmoral fashion. He has read newspaper descriptions of this thing. He recognises the fashion brought fresh from home. Trailing ends fringe the side of the skirt.

Perfect strangers do indeed talk to him about the incident. A number of gossips have worked themselves up into a state of indignation. What hysteria catches these stations of the plains! What pets these people get into!

As for the girl, is she sufficiently outside the bulk of cantonment society not to notice the difference between non-recognition and ostracism? She evinces a remarkably indifferent attitude.

"You look very dashing tonight, Miss Urquhart."

She does not simper. Nor does she thank him gushingly.

भारत के सब प्रदेशों का हर प्रकार का शुद्ध रेशमी कपड़ा।
ऊनी कपड़ा
तैयार सूती और ऊनी कपड़े—बच्चों के पुरुषों के और स्त्रियों के।
पशमीना शाल कपड़ा और साड़ियाँ।
सब प्रकार की दक्षिणी वेष-भूषायें।
भारतीय कशीदे की चीजें।
राष्ट्र ध्वज।
चरखा और सरजाम।
अहिंसक शहद।
कश्मीरी केशर
अहिंसक चमड़े की चीजें।
गांधी साहित्य।
भारतीय कला और संस्कृति साहित्य।

हाथ से बनी ग्रामोद्योग की वस्तुओं को
की और हाथ से बुनी प्रमाणित खादी का

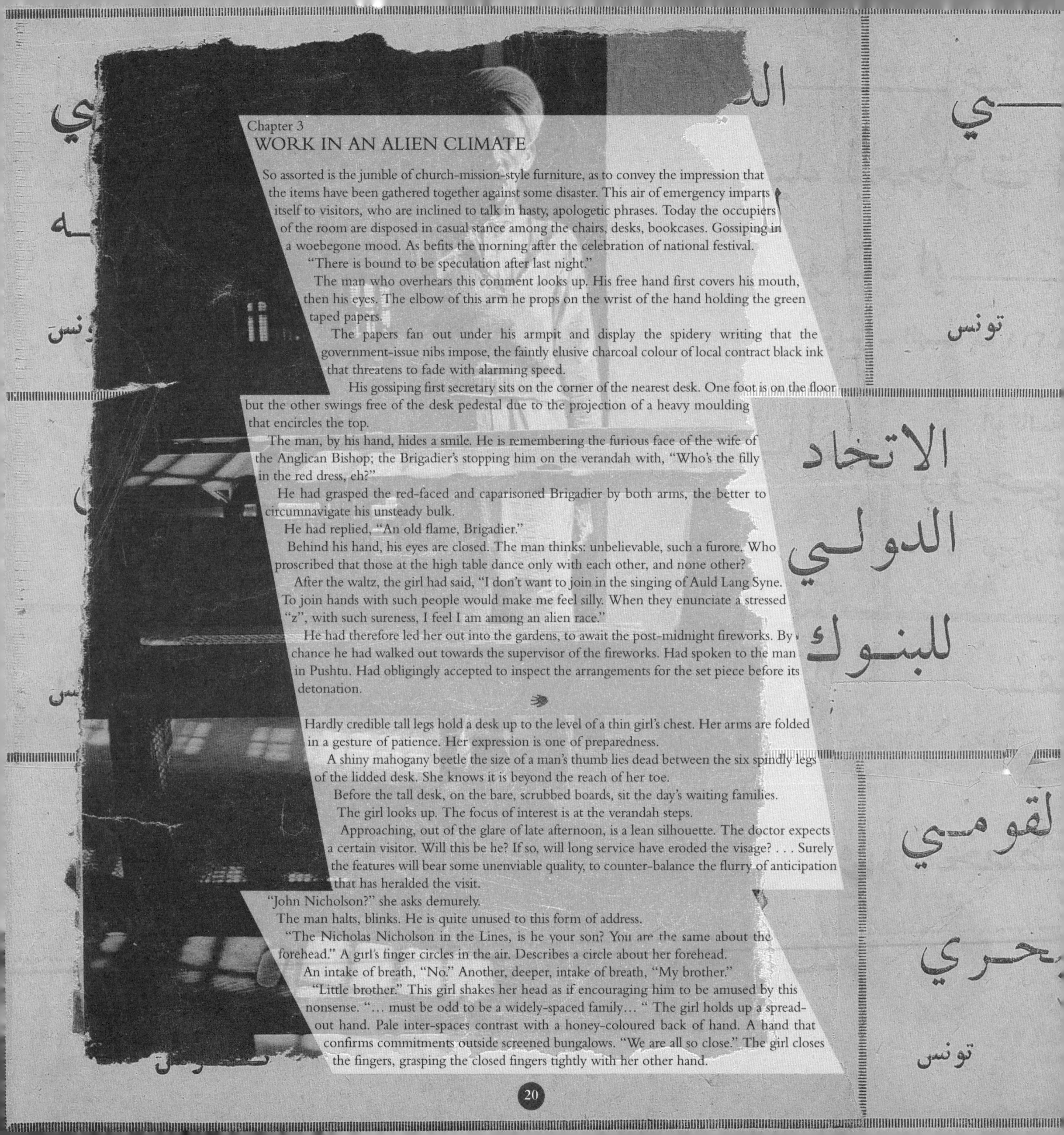

Chapter 3

WORK IN AN ALIEN CLIMATE

So assorted is the jumble of church-mission-style furniture, as to convey the impression that the items have been gathered together against some disaster. This air of emergency imparts itself to visitors, who are inclined to talk in hasty, apologetic phrases. Today the occupiers of the room are disposed in casual stance among the chairs, desks, bookcases. Gossiping in a woebegone mood. As befits the morning after the celebration of national festival.

"There is bound to be speculation after last night."

The man who overhears this comment looks up. His free hand first covers his mouth, then his eyes. The elbow of this arm he props on the wrist of the hand holding the green taped papers.

The papers fan out under his armpit and display the spidery writing that the government-issue nibs impose, the faintly elusive charcoal colour of local contract black ink that threatens to fade with alarming speed.

His gossiping first secretary sits on the corner of the nearest desk. One foot is on the floor but the other swings free of the desk pedestal due to the projection of a heavy moulding that encircles the top.

The man, by his hand, hides a smile. He is remembering the furious face of the wife of the Anglican Bishop; the Brigadier's stopping him on the verandah with, "Who's the filly in the red dress, eh?"

He had grasped the red-faced and caparisoned Brigadier by both arms, the better to circumnavigate his unsteady bulk.

He had replied, "An old flame, Brigadier."

Behind his hand, his eyes are closed. The man thinks: unbelievable, such a furore. Who proscribed that those at the high table dance only with each other, and none other?

After the waltz, the girl had said, "I don't want to join in the singing of Auld Lang Syne. To join hands with such people would make me feel silly. When they enunciate a stressed "z", with such sureness, I feel I am among an alien race."

He had therefore led her out into the gardens, to await the post-midnight fireworks. By chance he had walked out towards the supervisor of the fireworks. Had spoken to the man in Pushtu. Had obligingly accepted to inspect the arrangements for the set piece before its detonation.

Hardly credible tall legs hold a desk up to the level of a thin girl's chest. Her arms are folded in a gesture of patience. Her expression is one of preparedness.

A shiny mahogany beetle the size of a man's thumb lies dead between the six spindly legs of the lidded desk. She knows it is beyond the reach of her toe.

Before the tall desk, on the bare, scrubbed boards, sit the day's waiting families.

The girl looks up. The focus of interest is at the verandah steps.

Approaching, out of the glare of late afternoon, is a lean silhouette. The doctor expects a certain visitor. Will this be he? If so, will long service have eroded the visage? . . . Surely the features will bear some unenviable quality, to counter-balance the flurry of anticipation that has heralded the visit.

"John Nicholson?" she asks demurely.

The man halts, blinks. He is quite unused to this form of address.

"The Nicholas Nicholson in the Lines, is he your son? You are the same about the forehead." A girl's finger circles in the air. Describes a circle about her forehead.

An intake of breath, "No." Another, deeper, intake of breath, "My brother."

"Little brother." This girl shakes her head as if encouraging him to be amused by this nonsense. "… must be odd to be a widely-spaced family… " The girl holds up a spread-out hand. Pale inter-spaces contrast with a honey-coloured back of hand. A hand that confirms commitments outside screened bungalows. "We are all so close." The girl closes the fingers, grasping the closed fingers tightly with her other hand.

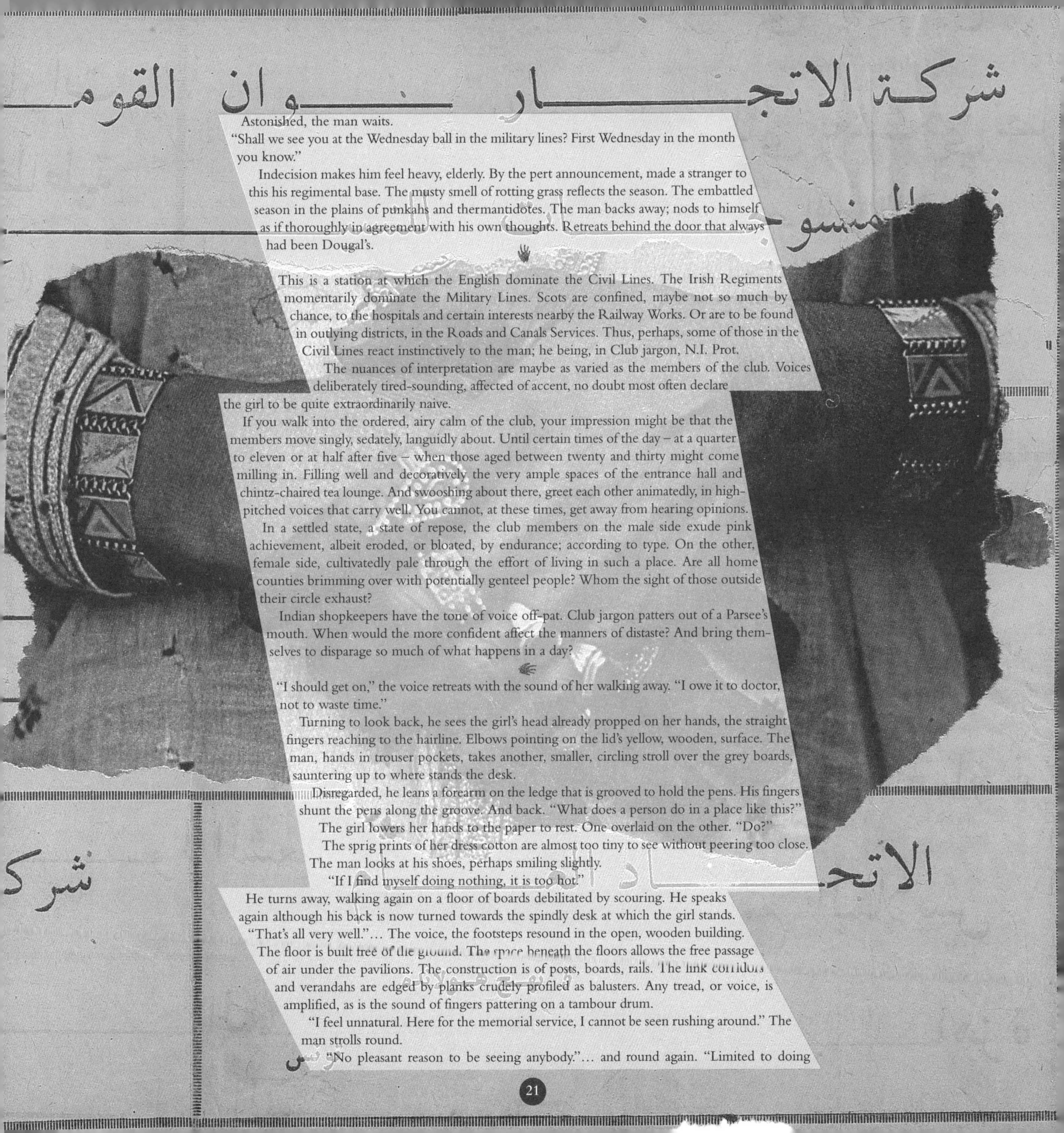

Astonished, the man waits.

"Shall we see you at the Wednesday ball in the military lines? First Wednesday in the month you know."

Indecision makes him feel heavy, elderly. By the pert announcement, made a stranger to this his regimental base. The musty smell of rotting grass reflects the season. The embattled season in the plains of punkahs and thermantidotes. The man backs away; nods to himself as if thoroughly in agreement with his own thoughts. Retreats behind the door that always had been Dougal's.

This is a station at which the English dominate the Civil Lines. The Irish Regiments momentarily dominate the Military Lines. Scots are confined, maybe not so much by chance, to the hospitals and certain interests nearby the Railway Works. Or are to be found in outlying districts, in the Roads and Canals Services. Thus, perhaps, some of those in the Civil Lines react instinctively to the man; he being, in Club jargon, N.I. Prot.

The nuances of interpretation are maybe as varied as the members of the club. Voices deliberately tired-sounding, affected of accent, no doubt most often declare the girl to be quite extraordinarily naive.

If you walk into the ordered, airy calm of the club, your impression might be that the members move singly, sedately, languidly about. Until certain times of the day – at a quarter to eleven or at half after five – when those aged between twenty and thirty might come milling in. Filling well and decoratively the very ample spaces of the entrance hall and chintz-chaired tea lounge. And swooshing about there, greet each other animatedly, in high-pitched voices that carry well. You cannot, at these times, get away from hearing opinions.

In a settled state, a state of repose, the club members on the male side exude pink achievement, albeit eroded, or bloated, by endurance; according to type. On the other, female side, cultivatedly pale through the effort of living in such a place. Are all home counties brimming over with potentially genteel people? Whom the sight of those outside their circle exhaust?

Indian shopkeepers have the tone of voice off-pat. Club jargon patters out of a Parsee's mouth. When would the more confident affect the manners of distaste? And bring themselves to disparage so much of what happens in a day?

"I should get on," the voice retreats with the sound of her walking away. "I owe it to doctor, not to waste time."

Turning to look back, he sees the girl's head already propped on her hands, the straight fingers reaching to the hairline. Elbows pointing on the lid's yellow, wooden, surface. The man, hands in trouser pockets, takes another, smaller, circling stroll over the grey boards, sauntering up to where stands the desk.

Disregarded, he leans a forearm on the ledge that is grooved to hold the pens. His fingers shunt the pens along the groove. And back. "What does a person do in a place like this?"

The girl lowers her hands to the paper to rest. One overlaid on the other. "Do?"

The sprig prints of her dress cotton are almost too tiny to see without peering too close. The man looks at his shoes, perhaps smiling slightly.

"If I find myself doing nothing, it is too hot."

He turns away, walking again on a floor of boards debilitated by scouring. He speaks again although his back is now turned towards the spindly desk at which the girl stands.

"That's all very well."… The voice, the footsteps resound in the open, wooden building. The floor is built free of the ground. The space beneath the floors allows the free passage of air under the pavilions. The construction is of posts, boards, rails. The link corridors and verandahs are edged by planks crudely profiled as balusters. Any tread, or voice, is amplified, as is the sound of fingers pattering on a tambour drum.

"I feel unnatural. Here for the memorial service, I cannot be seen rushing around." The man strolls round.

"No pleasant reason to be seeing anybody."… and round again. "Limited to doing unpleasant things. From which there is no escape." He pauses, returned from this last circling stroll. And looks directly at his silent compatriot. Her mouth turns up at the corners, in a madonna's silly smile. "What are you thinking?" He asks this rather sharply.

The girl ducks her head, touching the papers as if indeed conscience stricken.

He barely hears her evasive reply.

"If I had not wanted to know I would not have asked, now would I?"

However, she considers further, before telling.

"You might as well get on with your reading," is all he can think of saying. "You cannot have listened to anything I said, all afternoon."

She protests. And follows the man as he walks away from her. The shoulders inside the pepper-and-salt suit heave a shrug.

Then the girl hears the orderlies opening the doors at the end of the open corridor. This means the doctor will soon walk through. She ceases to follow the man. Instead she darts back to her desk.

The man hears her move to do so before he hears her blurt out, "I must get on, I shall not be ready."

Minto Road proves to be a tiny cul-de-sac, off Hardinge Drive. One of three narrow, dissimilar strips of hogging. The site is a triangular piece of ground, formed as the Drive and the Parade converge.

A tinkle of carriage harness sounds close in its passing either side. Carriages sound as if they move restlessly; towards and from the Grand Parade before the Military Lines. Distant flares light the haze that drifts about the Drive at its more posh end. The smell of Hoolie fireworks hangs in the smoke that permeates the air.

In the cul-de-sac, shutters shrunken by the heat show chinks of yellow light. The blades of light scythe the darkness. Light edges glint in the smoke-touched air as if of that instant cutting through the air.

The wood-built bungalows are painted blue. A chalky native blue. The bungalows are of the type bachelors share to economise on servants. Number 2 is situated on the corner. With a frontage, behind an extra high wall, facing on to the Drive.

From the crude portico in the wall, a gate lets into a pergola wild with the tendrils of a creeping plant. The path leads to steps, which in turn lead to a passage. Doors lacking, this passage opens into a day room. The second day room, just visible beyond, will be back to back with the accommodation of the other occupancy.

The two expecting their visitor are found sitting, either side of a fireplace. They sit over a token fire, lit to dry the air.

As he is shown through, she gets up saying, "My brother." And holds out her hand towards the youth beside her. This indeed surprises the visitor. Here stands a lean, pale, serious youth, who looks barely her senior, and probably is aged eighteen.

When they meet at the beginning of the evening she smiles brightly, "I hope we did not bore you."

"When? Oh. No. Why should you?"

"I don't know."

A hand rises to pinch the bridge of his nose. "I enjoyed myself." Which is the truth. "Did you not?"

"You did not shake hands."

With his laughter, conversation re-focuses. Thus she loses him again to other people in the room.

The servant who precedes them into the room does so very consciously. Legs waddling very slowly. Seeming independent of the body. Legs moving below a waist wrapped in a quantity of purple sash.

Yet the sirpesh at the side of the rigidly-upheld turban has a purple dyed hair fringe that vibrates with the waddle.

The girl feels as full of almost irrepressible giggles as gassified liquid in a glass marble-stoppered bottle.

Barely allowing an introduction to be made, the men turn back immediately to the problem of the rains undermining the newer portion of the Anglican graveyard. All evening, the same supercilious, impatient attitude of her compatriots. The black eyes of tray-bearing servants are the only ones to look at her as if she has any substance at all.

The end of the evening comes only after an interminable time. The outside lamp has almost filled up with insects while she has been inside. As she watches, a cloud of insects still fight to get in to join their trapped kind. Winged beetles audibly tapping the glass with heavy armoured bodies.

Too late she manoeuvres for a chance to speak. The last she sees of him is the back of the head and shoulders, driven away as the next carriage pulls up under the porte-cochère. As soon as she is in bed, she cries.

Unlike Minto Road, Company Drive, renamed Imperial Drive in '60, is quite an unused name except in directions to delivery carts: the cognoscenti needed no more than the individual names of Magnolia, or whatever, to locate further the imposing villa dwellings.

The distance across the cantonment is so great, and she so impatient for the visit and what should be said, that only in retrospect will come any awareness of the porte-cochère and the wide uprise of marble steps; of the spacious half-landing outside the lodge, of the door servants, the choukedar standing at attention; the final flight, the long open corridor.

On the floor is thick cotton cloth secured by hitching brass eyelets over brass studs set in the marble. The servant precedes the girl, the corridor passes the edge of two open day rooms. The foot of a wide internal stair, whose intermediate landing is so high its surface, although well lit by a lantern light, is out of sight to her. It is said, rich houses have chintz-lined ceilings to their bedrooms, some women lying under sprigged and pleated muslin tents. Maybe the rooms above have such gracious ceilings. The corridor continues, between the walls of rooms with large, closed, mahogany-panelled doors.

The farthest verandah proves to be empty.

Deserted by the servant, the girl experiences a sense of being stranded. Left sitting facing a wide spread circle of wicker chairs in such poor condition that perhaps no one comes to this verandah. Silence reigns in the house behind her. Prolonged use has splayed the bases of the chairs, given each a tilt out of plumb that is disconcerting to look at. The girl, made uncomfortable by such dilapidated furniture, involuntarily straightens her back.

Each cushion is not only so worn as to show its weaving through the print of its pattern, but is also soiled. Chintz long since bleached from recognisable flowers; jig-saw remnants of a certain deep red are now the pattern. All she surveys adds horror to her being left alone.

The girl has chosen to sit on one of the couches, in the belief that when the man enters, he will seize the opportunity to sit beside her.

The company to take tea assembles slowly. As if taking tea bores them. Entering, nodding or mumbling to her being there, they seem to continue a reverie among themselves. The servants silently, simultaneously appear and stand with their utensils. The khansama holds a small silver tray bearing small silver jugs and bowls. He stares straight ahead. On the low table, moved up and placed in front of her, the servants set down a vast tray with an upstand rim. A rim finely perforated; it is like lace. She had not anticipated being expected to pour. Two silver pots match the two tall jugs. Ivory pads are riveted between their handles and their bodies.

One of the secretaries leans forward. With his hand he twists one of the pots half round. Without letting go his hold, he revolves the pot in reverse. The face at first towards her is visible again. A pointing finger indicates a rococo frame engraved on the bellied side of the pots: within the etched maidenhair fern borders can be read rustic lettering denoting one

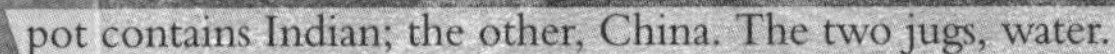

pot contains Indian; the other, China. The two jugs, water.

She pours as steadily as she can. Uses both hands, as she had seen her grandmother need to do. So scant her practice with such heavy pots, the girl does not dare look up at the sound of voices: those late joining the company at tea, now enter.

An under-servant, murmuring only, "Three for China, memsahib," lifts each cup filled on to the tray the khansama holds. Each cup is presented by this servant to each recipient.

Last of all, the khansama offers his tray to her. The girl grasps the fatter of the jugs. Pouring, she recognises goat's milk. The other jug will contain boiled cow's milk. She returns the lightweight jug shakily to the tray. A turn of her head indicates refusal of the crystal and the loaf sugar. Her hand is raised against the offering by a servant of the silver covered dish. When she refuses its offering she does not know what its hinged lid might cover. Her concentration has been on pouring. Those at company are eating thin bars of buttered toast. She does not wish to make a noise eating toast.

The space beside her remains untaken.

A replenishment of cups with tea is desired by all persons save herself.

A wall of foliage leans close by the verandah rail. Its knotty growth suggests it has been planted by some previous occupant of the Residency. Its closeness to the rail encourages a sense of ennui. Crowding has caused the branches to have grown attenuated. The thin branches weakly uphold too few heat-furled leaves. The too long stems are unattractive. The dusty leaves hang like discarded dinner napkins.

At last, the assembled company rise to their feet. This movement declares tea over. The dispersal begins hesitatingly.

The last mahogany door closes, shuts away the inconsiderate.

The girl is aware enough now of the nature of her surroundings to note the embrasures on either side also open on to sun-smitten grey branches. Further along the corridor, crowding shrubs of lanky vegetation are on the right-hand side, growing in a courtyard. The girl, while she waits, has stepped aside into one of these recessed spaces.

Is now resting a clenched fist on the stone capping of a parapet. She slowly rubs a left hand up, and down, the out-reaching arm. The shape of fashion conceals the slenderness of the arm.

There has been an ample elapse of time for her to decide what to say. But she has been unable to muster thought of how to protest at the manner of her reception. Not only this lack of presence of mind. All pre-thought argument concerning the drama of previous situations has slipped out of words.

The man walks towards the embrasure. His feet tread quietly on the canvas runner.

At a loss for words, obedient, she allows herself to be led out… stair… landing… stair… The door servants are bowing outside their lodge. Too late, she realises there is no greater privacy. Now on the carriage steps, any chance of an exchange has been missed.

A man is sleeping on his side. On a tired cloth, spread out. On the white marble paving of the great courtyard. Does he work here? Has he been sweeping the yellowing, marble squares during the hottest part of the day and therefore now has need of rest?

Other men sit about. One mending a sandal. Indians use their monuments so casually – she thinks – as places just to do odd tasks in.

As they come out, a man on crutches holds out an indented brass begging bowl. How can crutches get so worn in the shaft for the wood to become threadbare and need bandaging in rags? Only one child ignores the admonitions of the carriage servant. The wisp of brown child picks something out of the filth. The carriage regained, she sits forward. This is to allow the air movement created by the forward passage of the carriage to flow between her back and the warmth of the carriage upholstery. Her head is bowed, as if looking at her hands, turned palm upwards in her lap. She breathes slowly, deeply; glad the sabbath is nearly over.

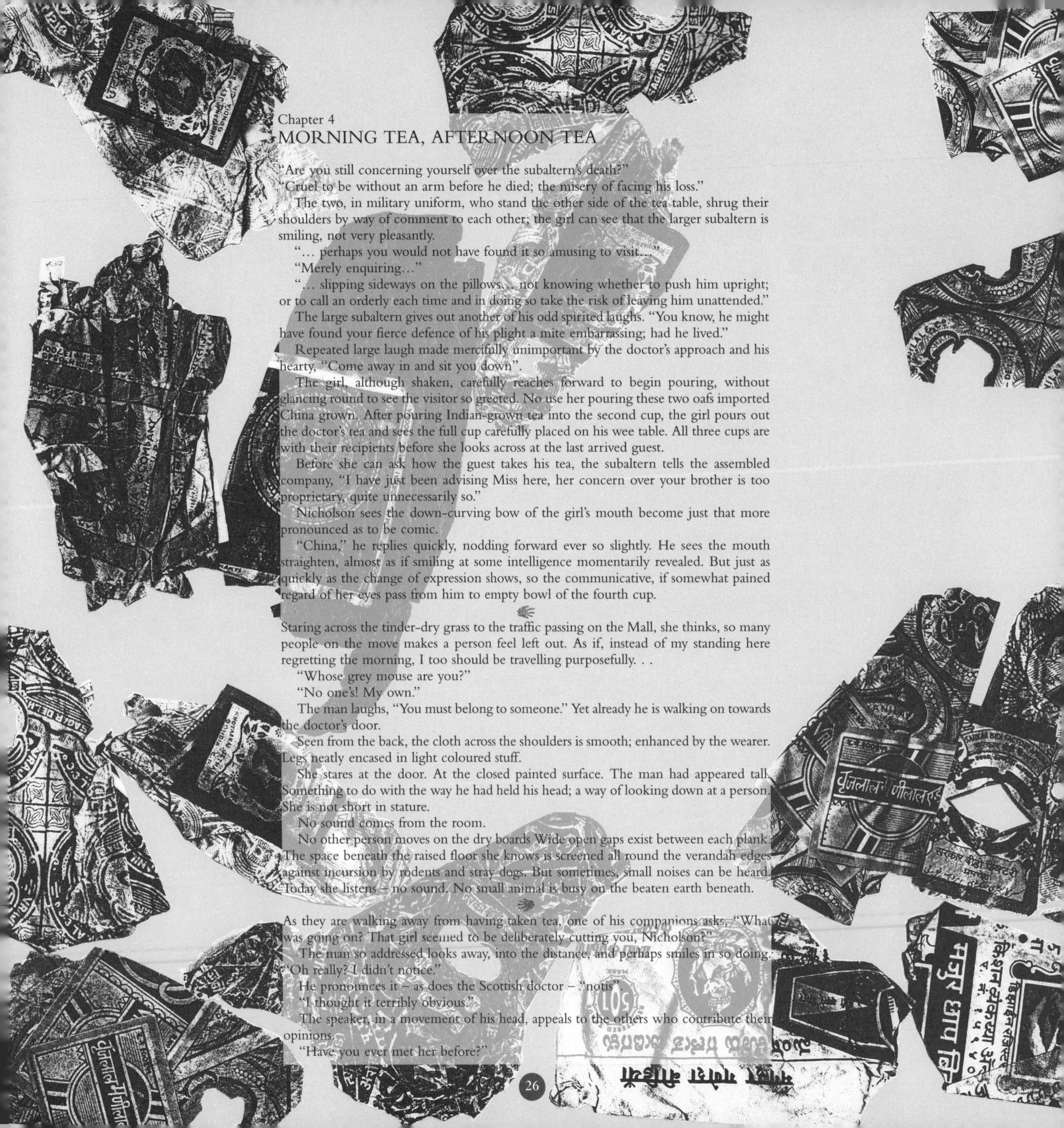

Chapter 4

MORNING TEA, AFTERNOON TEA

"Are you still concerning yourself over the subaltern's death?"

"Cruel to be without an arm before he died; the misery of facing his loss."

The two, in military uniform, who stand the other side of the tea table, shrug their shoulders by way of comment to each other; the girl can see that the larger subaltern is smiling, not very pleasantly.

"… perhaps you would not have found it so amusing to visit…"

"Merely enquiring…"

"… slipping sideways on the pillows… not knowing whether to push him upright; or to call an orderly each time and in doing so take the risk of leaving him unattended."

The large subaltern gives out another of his odd spirited laughs. "You know, he might have found your fierce defence of his plight a mite embarrassing; had he lived."

Repeated large laugh made mercifully unimportant by the doctor's approach and his hearty, "Come away in and sit you down".

The girl, although shaken, carefully reaches forward to begin pouring, without glancing round to see the visitor so greeted. No use her pouring these two oafs imported China grown. After pouring Indian-grown tea into the second cup, the girl pours out the doctor's tea and sees the full cup carefully placed on his wee table. All three cups are with their recipients before she looks across at the last arrived guest.

Before she can ask how the guest takes his tea, the subaltern tells the assembled company, "I have just been advising Miss here, her concern over your brother is too proprietary, quite unnecessarily so."

Nicholson sees the down-curving bow of the girl's mouth become just that more pronounced as to be comic.

"China," he replies quickly, nodding forward ever so slightly. He sees the mouth straighten, almost as if smiling at some intelligence momentarily revealed. But just as quickly as the change of expression shows, so the communicative, if somewhat pained regard of her eyes pass from him to empty bowl of the fourth cup.

Staring across the tinder-dry grass to the traffic passing on the Mall, she thinks, so many people on the move makes a person feel left out. As if, instead of my standing here regretting the morning, I too should be travelling purposefully. . .

"Whose grey mouse are you?"

"No one's! My own."

The man laughs, "You must belong to someone." Yet already he is walking on towards the doctor's door.

Seen from the back, the cloth across the shoulders is smooth; enhanced by the wearer. Legs neatly encased in light coloured stuff.

She stares at the door. At the closed painted surface. The man had appeared tall. Something to do with the way he had held his head; a way of looking down at a person. She is not short in stature.

No sound comes from the room.

No other person moves on the dry boards Wide open gaps exist between each plank. The space beneath the raised floor she knows is screened all round the verandah edges against incursion by rodents and stray dogs. But sometimes, small noises can be heard. Today she listens – no sound. No small animal is busy on the beaten earth beneath.

As they are walking away from having taken tea, one of his companions asks, "What was going on? That girl seemed to be deliberately cutting you, Nicholson?"

The man so addressed looks away, into the distance, and perhaps smiles in so doing. "Oh really? I didn't notice."

He pronounces it – as does the Scottish doctor – "notis".

"I thought it terribly obvious."

The speaker, in a movement of his head, appeals to the others who contribute their opinions.

"Have you ever met her before?"

"What was she playing at?"

"I really didn't notice."

The four men walk on. The tamarind trees cast too little shade. The squirrels seem fewer in number, or less active than usual; yet, no doubt, they will materialise in half an hour or so.

Thursday sees the start of a week of local festivals which make but slight difference to the number of itinerants attempting sales at the hospital windows.

Since the dispensary adjoins the hospital compound, those on the dispensary verandah can see from afar these predators… scrawny, gay children; begging women with outstretched arms; men, thin and fat. Bird sellers are the most encumbered; birds being considered suitable to amuse the sick… but who can imagine that mangoes, bananas, opened native fruit, fried meat pieces on split cane – all proffered to the window sills on long wands – will benefit the sick of any regiment?

The girl watches from afar. Smooths her hand slowly, tenderly up her arm: why is the arm top considered so efficacious a place to receive the new treatment?

She looks at the faces of those seated, taking tea; are any of these minds around her a gabble of thoughts, and dreams, as unconnected with tea as are hers?…

Why am I here if nothing worthwhile is to happen to me?…

What is going to be the achievement of my life if I don't marry?

Thinking so, consumed by her own wanting, she is reminded of the burning bush.

It is so hot that she keeps thinking… oh! I am so tired… yet there is no reason to be; she is not carrying a doolie, nor picking at the hard earth for a drain from the latrines; not scraping a bunch of twigs across a verandah; people said they were lazy! Even dressed in the finest of thin muslins, just bending down to pick up a fallen flower is enough to flush her cheek. These thoughts are interrupted as her brother joins those already seated on the verandah.

"Oh! what a day! I thought it would never end; our punkah boys just managed to make the air feel more and more like our being stirred in syrup; it's got to break soon."

Verandah tea again; the girl is sitting so as to present her back to Nicholson who is ideally stretched out for ostracism in the deep basket chair which is only comfortable if its foot-rest is hooked on. The two elderly gentlemen taking tea with the doctor therefore form a tight enclave for whom she pours. First two cups pass to the old India hands who, one after the other, reach out both hands to take off the tray their cups; as if these are weighty objects. Their wrists are crooked to this task which proceeds with deliberation. The full cups are lowered onto the wide splay arms of the basket chairs. The liquid in the cups is then fully in the view of the recipient. The elderly gentlemen supervise the kitmutgar as he puts in their tea much, much, sugar; just a soupçon of milk. Hawk-eyed, beetle-browed, each gentleman in turn watches the tea being properly stirred. All the sugar has to be dissolved; the one thing they cannot abide is to find grains of the coarse local sugar in the dregs. The doctor prefers the milk and sugar put in the cup prior to the tea being poured; the kitmutgar knows to hold his tray towards the extra large moustache cup. The doctor is the third to receive his tea.

Then she motions to the servant to bear the remaining guest's cup on his brass tray to the sahib seated slightly behind her back.

The doctor sits stirring his tea, his knees spread wide to engulf the little tamourine table. The cup is placed well inside the knobbly rim, plaited with a strand of black dyed grass.

The girl herself offers the biscuit plate to the doctor; he takes a plain and a sweetened biscuit in between fingers of his left hand, deftly, almost without glancing at the two parallel rows of biscuits laid out on the linen mat on the plate. He will nibble them so that he tastes neither unadulterated by the other.

The kitmutgar reappears behind the doctor and – as on the day before – the girl signals he is to offer the biscuits to the sahib behind. Handing the brass tray to the boy to hold, in a steady turning movement so as not to tinkle the metal jugs together, the kitmutgar moves to grasp the biscuit plate; but the guest stops him, speaking Hindi. The kitmutgar is still bent, replacing the biscuit plate on the table; he pauses in that position at the abrupt question.

"What is going on? What kind of behaviour is this?"

The silence lasts until Mr Bingham, in picking up his cup, spills some tea in the saucer and half angrily says something, either in agreement, or by way of excuse as he empties tea from his saucer on to the boards.

The girl feels half annoyed and half wanting to laugh that her petty manoeuvres have been noticed. Mr Nicholson has every right to make her appear stupid.

"I don't think there is anything, Dougal."

"Why is Clare not being civil to you? Clare? What is this?"

"What have I done to upset you, Doctor?"

"I'm sure I am not the only one to notice. Such silly behaviour."

Mr Bingham, in a movement this time, definitely assents.

Mr Fergusson clicks his china teeth.

"If this is some sort of party game, we ought to know it."

"It's too bad," mumbles Mr Fergusson.

Mr Bingham's chair creaks. The girl sits straighter, her head bent, lips pursed as if otherwise a smile might break out.

"These young girls ought not to be away from parental control," Mr Bingham says to the air at his side. "Girls need parental control."

Mr Fergusson manages both to nod agreement and finish drinking his tea at the same time. Mr Fergusson lets his cup be taken. He watches, as the dregs are emptied out into the basin, the cup replenished, and returned. Small sounds of tea time; china gently touching, cup wobbling in saucer… as a background, sound coming from the old city like a dragon's breath, builds up beyond the bird racket; the people again begin to move about; from one end of the city to the other.

"It's too bad." The doctor gives his tea a final cooling stir, lays down the teaspoon in the saucer, lifts the cup slowly in both hands, stooping forwards slightly as he takes his first sip. "I don't ever remember Lettie doing anything at all childish, you know, Clare."… a second sip… "It's too bad."

"We must get on, soon," Mr Fergusson says to Mr Bingham.

As he comes to join those at tea on the verandah, the boy is dispatched to fetch more cups. The doctor turns and calls after the small figure – more white cloth than boy – "And another pot of China", this instruction in Urdu.

Without a pause the man walks towards the remaining empty chair. However, during the same moments as the man is approaching, and the boy departing, by a smartly delivered thrust on the front edge of the chair's seat, the girl puts the chair out of the circle. The immediate re-positioning of her own chair, by grasping the sides of the seat, prevents the replacing of the empty chair within the circle.

In the muddle of the dispensary server stepping forward just as the man does towards the vacant chair, the girl reaches out to the rim of the table. Other arms are also outstretched. The man has his attention drawn back to the girl by the sound of the table legs scraping over the boards: her whole attitude is provocative in some way he cannot define.

The man is seated in the chair, placed in the meagre space created by other guests' chairs shifting sideways. He pours milk in his cup, replacing the jug thoughtfully. Then leans back, to indicate to the boy he can take his tray elsewhere. As the man drinks his tea, he finds he is in a position to scrutinise those at tea more so than if he had been fully of the circle.

"Give our guest more tea, now."

As reply to the doctor, the girl directs the boy to where Nicholson sits; and wishes she had thought to pretend not to know which guest had been meant: she can save that ruse for next time.

The visitor collaborates by proffering his cup to the kitmutgar… the girl receives the cup… the cup is returned by the same hands. Nothing is wrong in all this; all actions are normal enough, on the surface.

No one says anything. Indeed, conversation lapses during the emptying and replenishing of the cup. The cup already tipped to empty out the dregs, a further elaboration of insult occurs to her: next time this can be forgotten due to attention to what a neighbour is saying.

Morning and afternoon tea game. To be played in May, when woven screens still smell of grass.

He watches her skirt brushing against the legs of the furniture; the room had not struck him as miniscule but the skirt fills the space from bureau leg to table leg, from cabinet front to chair leg. Why does she think inconsequential chit-chat requires such an effort on her part? Slowly gathering his wits, he attempts to ease her out of her self-inflicted embarrassment. This requires a renaissance of his youth's skittish humour: this with some difficulty. A polished wit that shines in the mess not only becomes deliberately dulled outside its walls, but the communality of words and events cannot exist outside. Besides, since years he has adopted a manner designed to withstand the expatriate female trying to marry-off some undesirable maiden. The man watches the green and blue stripes at the hem, mesmerised by the movement.

Listening to the skirt, he raises his hand in thought to the parchment-like skin on the bridge of his nose. He has the uncanny sense of watching from the wings a play meant just for him, but to whose theatre he has been called by some ruse.

The emotion winging through the air can literally be cut with a knife… He is in the thick of this as soon as he walks on to the verandah at tea-time. Even if the dry atmosphere is taken into account, never before in the plains has he experienced such a sparked aura as around the girl… no doubt, emanating from this one person… Some choice is being willed into the making, neither bad, nor snobbish… there are none of the usual hostess's graduated nuances of regard towards each of her assembled guests. Not the usual station insularity shown the outsider. What is being perpetrated is not at all subtle. But most militant. Is this a pass at him?… He almost chokes on a throat full of tea to suggest to himself such a crude approach.

"I have been meaning to enquire if you knew that the subaltern Nicholas was my brother?"

Laughter… totally unexpected; "I have been in a turmoil, as to whether… or… "

"Or?" he prompts, his fingers shaking slightly as they rise to the bridge of his nose, where they rest for a moment.

The girl places her hands boyishly on the verandah rail behind her hips, draws in a laughing breath. She speaks in such a whisper that he leans forward to hear… then hearing, steps back.

"That I was older than my brother made him young, not, ipso facto, that I was his father."

Down-turned bow of the girl's mouth melts. She says sweetly, "Why have you never married, Mr Nicholson?"

"The canal has barely opened, Miss Urquhart. There have never been so many women in India who were not either married or widowed; I assume you are in neither of these categories."

The down-turned bow mouth reforms, and twists away in a smile again; she says with a gurgle in her voice, "Then I am a very lucky girl; don't you agree, Mr Nicholson?"

"In what way fortunate?"

But she laughs at him and swings away; sits down, flattening her skirts behind her with her hands first, and starts to pour out tea: for a moment the surroundings seem unreal to him; as if this roguish incident is dreamt.

Morning tea is taken in the doctor's room because at this season the sun glances the dispensary verandah until midday. Morning callers sometimes join the doctor in this confined space; but more usual, at the start of the late afternoon, when they can dispose themselves on the verandah.

There might be military surgeons from the Fort; or the new surgeon from the Subscribers' Hospital compound beyond their wall; perhaps the two supervisors of male nurses; or someone come through from visiting the sick.

Alerted this morning by the small gong, the girl turns round and starts walking towards the door. She can see up the wide, open corridor that leads towards the hospital. Doctor stands outside the other door to his room, talking to the Eurasian orderly. Opening the door facing her, she sees the tea tray – as announced by the gong – is in position on the corner of the desk. But also – for she had not known him to be in the building – towards her, the unmistakable long back belonging to Mr Nicholson. The visitor twists round, forearm resting on the cabinet top. So as not to show a trace of guilt, the girl walks forward until her fingers lightly touch the edge of the white cloth that covers the black lacquered papier-mâché tray. She waits.

However, the doctor looks into the room only to say, "I will be right back. Pour out the tea now, look after our guest."

As the door closes, the girl, hands before her in the attitude favoured by painters of Christ before Pontius Pilate, challenges, "I thought no one noticed the first tea time – I was not even sure you noticed – did you?"

The man shifts his weight on to his other leg, repositioning his arm resting on the cabinet as he does so. "Vaguely," and wonders what will come next. He finds this focus of attention strangely flattering, even relaxing… poor Dougal had not been able to see it at all.

The girl feels consternation prickle out of her skin at the awkward silence.

"I will never get used to young girls today not being as sensible as Nettie was."

The girl shrinks back in her chair which disobligingly creaks.

The doctor gazes over her head to a framed, mottled, pencil drawing hanging on the wall.

"Did you know her?" The girl first has to cough to clear her throat.

Nicholson watches the cup balanced in her lap – a feat that never fails to fascinate him – yet he knows who was spoken to and shakes his head in answer to the question.

"Know who?" the doctor demands. "To my mind, Clare, you never bother to form a proper sentence in your head before you speak."

"I wondered if Mr Nicholson knew your wife," the girl whispers hoarsely, then coughs again.

"Oh no, no, no." The doctor's tone is cutting.

"I had no idea you lost her so early on." Nicholson looks directly at the doctor.

The doctor finishes drinking his tea and then holds the cup up, tilting it, inspecting it as though some specimen. He cannot abide these hair cracks so common in inexpensive china cups: he will break the china on the edge of his desk if it has the forbidden, offending blemish. "We had three lovely years together though."

The girl – now very near tears – covers her mouth with her hand, and stares wide-eyed at Nicholson to see if he too thinks it pitiful – such a small proportion of shared time, to time vacant.

"In those days, we reckoned eighteen out of nineteen would die out here. You can see them in the graveyard, all lined up under their stones and railings, urns and cherubs of the Goan carvers. Climate blamed; but always a better survival among Company Writers; they did the same work as you before you were born, John. Work for Company nine o'clock to twelve, dinner two to five o'clock; then, depending on seniority, the Writers returned; dish of tea; constitutional; supper; six days a week. Whereas you got your cadet, paying calls and drinking claret all morning."

The man leans forward, rising out of his chair to replace tea cup on tray, so his head approaches to half the previous distance from the girl and she can notice the way his eyelids crease, seem to imbue his eyes with confidence.

There follows the opening of the door; her walking back to her desk; the sound of following footsteps. In such heat, all motions tend to occur at a slow tempo.

She pauses and speaks before facing the visitor who she knows to be close behind her. "Will you come to afternoon tea?"

Nicholson chooses not to answer. He even turns away from the look of concern in her eyes. Instead he starts walking slowly out of the dispensary until he pauses on the second step down.

The girl, in her following him to the step above, is heard to do so by the visitor, who as a polite gesture, in order to face her, turns on the step.

Disliking to be faced with a man seemingly her size, the girl passes round him,

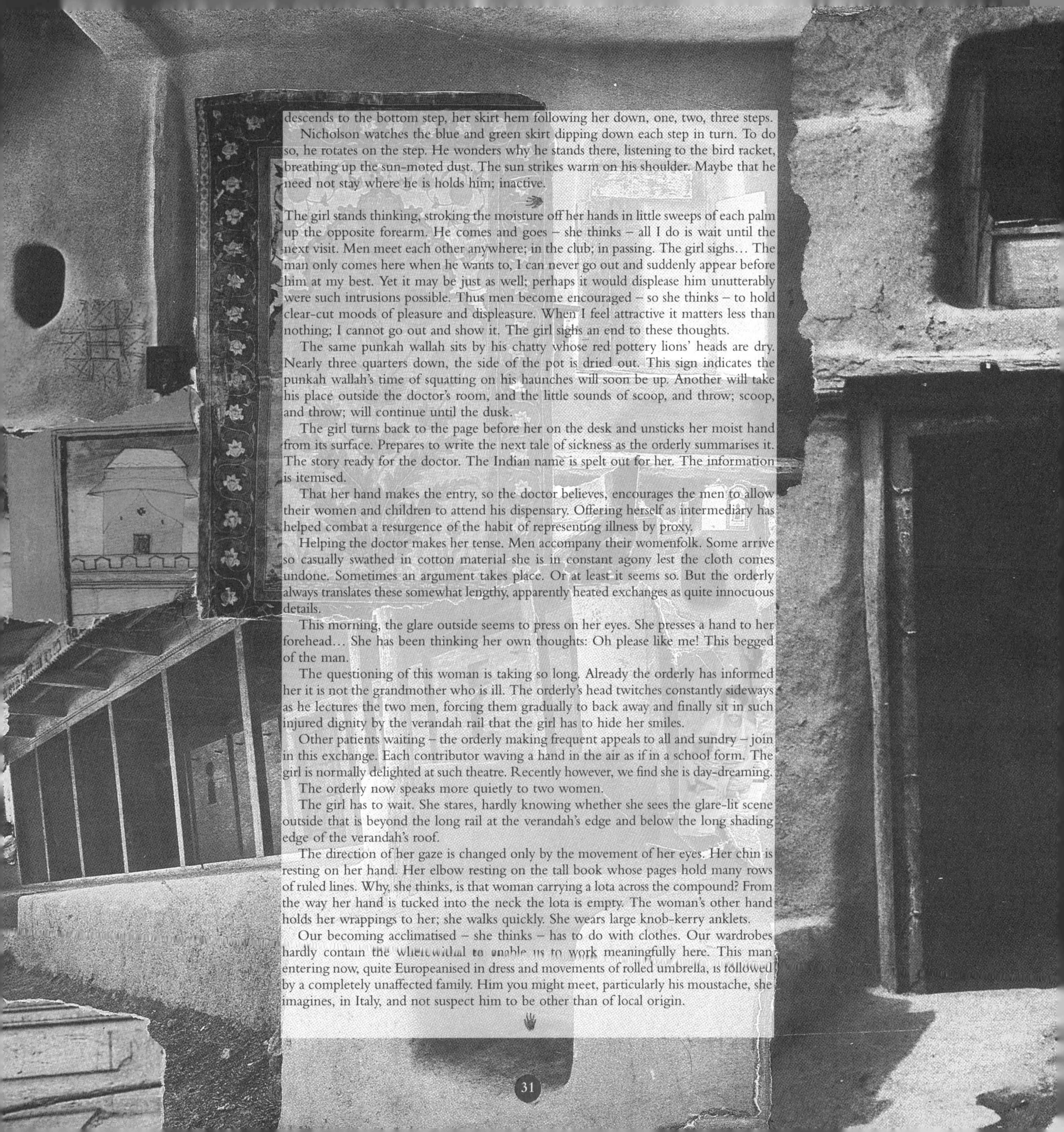

descends to the bottom step, her skirt hem following her down, one, two, three steps.

Nicholson watches the blue and green skirt dipping down each step in turn. To do so, he rotates on the step. He wonders why he stands there, listening to the bird racket, breathing up the sun-moted dust. The sun strikes warm on his shoulder. Maybe that he need not stay where he is holds him; inactive.

The girl stands thinking, stroking the moisture off her hands in little sweeps of each palm up the opposite forearm. He comes and goes – she thinks – all I do is wait until the next visit. Men meet each other anywhere; in the club; in passing. The girl sighs… The man only comes here when he wants to, I can never go out and suddenly appear before him at my best. Yet it may be just as well; perhaps it would displease him unutterably were such intrusions possible. Thus men become encouraged – so she thinks – to hold clear-cut moods of pleasure and displeasure. When I feel attractive it matters less than nothing; I cannot go out and show it. The girl sighs an end to these thoughts.

The same punkah wallah sits by his chatty whose red pottery lions' heads are dry. Nearly three quarters down, the side of the pot is dried out. This sign indicates the punkah wallah's time of squatting on his haunches will soon be up. Another will take his place outside the doctor's room, and the little sounds of scoop, and throw; scoop, and throw; will continue until the dusk.

The girl turns back to the page before her on the desk and unsticks her moist hand from its surface. Prepares to write the next tale of sickness as the orderly summarises it. The story ready for the doctor. The Indian name is spelt out for her. The information is itemised.

That her hand makes the entry, so the doctor believes, encourages the men to allow their women and children to attend his dispensary. Offering herself as intermediary has helped combat a resurgence of the habit of representing illness by proxy.

Helping the doctor makes her tense. Men accompany their womenfolk. Some arrive so casually swathed in cotton material she is in constant agony lest the cloth comes undone. Sometimes an argument takes place. Or at least it seems so. But the orderly always translates these somewhat lengthy, apparently heated exchanges as quite innocuous details.

This morning, the glare outside seems to press on her eyes. She presses a hand to her forehead… She has been thinking her own thoughts: Oh please like me! This begged of the man.

The questioning of this woman is taking so long. Already the orderly has informed her it is not the grandmother who is ill. The orderly's head twitches constantly sideways as he lectures the two men, forcing them gradually to back away and finally sit in such injured dignity by the verandah rail that the girl has to hide her smiles.

Other patients waiting – the orderly making frequent appeals to all and sundry – join in this exchange. Each contributor waving a hand in the air as if in a school form. The girl is normally delighted at such theatre. Recently however, we find she is day-dreaming.

The orderly now speaks more quietly to two women.

The girl has to wait. She stares, hardly knowing whether she sees the glare-lit scene outside that is beyond the long rail at the verandah's edge and below the long shading edge of the verandah's roof.

The direction of her gaze is changed only by the movement of her eyes. Her chin is resting on her hand. Her elbow resting on the tall book whose pages hold many rows of ruled lines. Why, she thinks, is that woman carrying a lota across the compound? From the way her hand is tucked into the neck the lota is empty. The woman's other hand holds her wrappings to her; she walks quickly. She wears large knob-kerry anklets.

Our becoming acclimatised – she thinks – has to do with clothes. Our wardrobes hardly contain the wherewithal to enable us to work meaningfully here. This man entering now, quite Europeanised in dress and movements of rolled umbrella, is followed by a completely unaffected family. Him you might meet, particularly his moustache, she imagines, in Italy, and not suspect him to be other than of local origin.

Chapter 5

LAYERS OF OCCUPATION

The sunset raises to iridescence the surface texture of the intervening ground. Any shade cast in the golden-apricot aura means that specks of gilt are withheld from certain particles of dust. This way, the diffuse shadows cast by the mango trees appear as paler, less luminescent areas. Obliquely striking light is made visible by foliage density; for the trees are huge. Their bulk, on this shadowed side towards the girl, is a dapple of grey blobs. The brightness of the lowering sun glints through a myriad apertures.

In the distance the Jamal al Shah is in silhouette; is bleached out of any solidity of form by the evanescent light: is small enough to be obscured by a raised hand.

Walls beyond the polo ground cast a shadow-band the self-same tint as their own surface shade. By this effect, all change of plane is lost.

"A grand sunset."

Without looking round, the girl smiles. She nods her head slightly, as if in agreement. The voice she hears sounds so even, perfect in pronunciation, as to be disembodied. As if the scene she admires gave tongue. The girl resists turning. Does not reply for fear that watching servants might find her talking to the air. Is it the heat she was not brought up to that makes her feel so strangely light of head?

The verandah she stands on has been coloured blue; although it is difficult now to be sure when coloured. The wicker chairs, once blue, hold traces of the paint in the underscoop of each weave. Where showing beneath the double layer of white tablecloth, the indentations in the wicker-bound legs hold the colour.

Beside the girl, hands outspread are rested on the verandah rail. Under one hand, lies a folder of papers. Without looking at the person of the hands, the girl speaks.

"Are sunsets always all-pervasive at this season?"

"Just before the weather breaks. They say the dust suspended in the air, the cause."

"The dust smells scented. I like it so. I suppose that sounds silly to anyone who is used to the season." Her shrug is expressive of a pleasure too great to bear. "Everything is so much more vivid than I imagined."

The man considers the lilt in her voice and judges her to be a Scot.

"I suppose you think that as I become more accustomed to this country, I shall cease to find the scene so decorative. But I am sure I shall not become disenchanted."

As she talks, with every slight movement, so shifts the gilding that falls on head and shoulders. Does he imagine that the rapt look is also fearful? Will it be a loss to miss a moment watching the sunset colours enrich… mutate?

The club servers are heard crossing the verandah. The girl turns slowly; as if loath to take her eyes off the sunset.

A line of tables is set against the back wall of the verandah. Without looking at the man standing at the verandah rail, the girl sits down at the now laid table… only then, as if returning to full awareness, asks the stranger, "Would you take some tea?"

Behind the man, and to one side, a body servant waits motionless, in a grey, buttoned-up cotton tunic.

Club servers stand in hierarchical order, in an aslant row at the verandah end. The colour of the sashes and sirpeshes of these make some sort of collective, compulsive movement. This slight movement means nothing to the girl. Until it is remembered at a much later time. Then, when after so long an interval remembered, it is unclear if meaningful movement had been intended.

There is a pause of some moments before there is an answer. The man so addressed across the verandah, turns. "I would like a cup."

As he sits, his stooping reveals that it is the body servant who moves and presents the chair. The stranger hands the folder to his body servant. Taped in the government

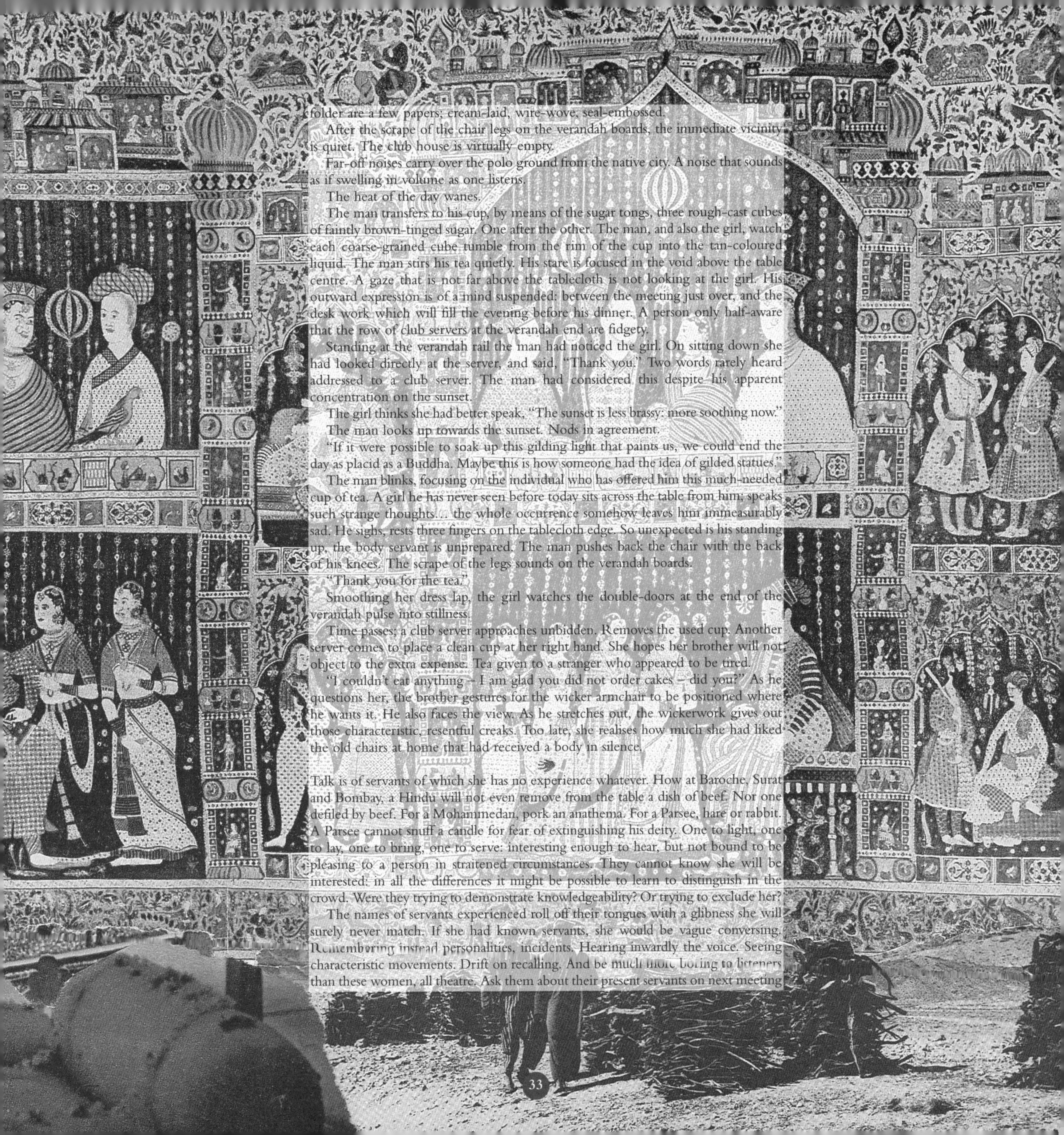

folder are a few papers; cream-laid, wire-wove, seal-embossed.

After the scrape of the chair legs on the verandah boards, the immediate vicinity is quiet. The club house is virtually empty.

Far-off noises carry over the polo ground from the native city. A noise that sounds as if swelling in volume as one listens.

The heat of the day wanes.

The man transfers to his cup, by means of the sugar tongs, three rough-cast cubes of faintly brown-tinged sugar. One after the other. The man, and also the girl, watch each coarse-grained cube tumble from the rim of the cup into the tan-coloured liquid. The man stirs his tea quietly. His stare is focused in the void above the table centre. A gaze that is not far above the tablecloth is not looking at the girl. His outward expression is of a mind suspended: between the meeting just over, and the desk work which will fill the evening before his dinner. A person only half-aware that the row of club servers at the verandah end are fidgety.

Standing at the verandah rail the man had noticed the girl. On sitting down she had looked directly at the server, and said, "Thank you." Two words rarely heard addressed to a club server. The man had considered this despite his apparent concentration on the sunset.

The girl thinks she had better speak, "The sunset is less brassy: more soothing now."

The man looks up towards the sunset. Nods in agreement.

"If it were possible to soak up this gilding light that paints us, we could end the day as placid as a Buddha. Maybe this is how someone had the idea of gilded statues."

The man blinks, focusing on the individual who has offered him this much-needed cup of tea. A girl he has never seen before today sits across the table from him; speaks such strange thoughts… the whole occurrence somehow leaves him immeasurably sad. He sighs, rests three fingers on the tablecloth edge. So unexpected is his standing up, the body servant is unprepared. The man pushes back the chair with the back of his knees. The scrape of the legs sounds on the verandah boards.

"Thank you for the tea."

Smoothing her dress lap, the girl watches the double-doors at the end of the verandah pulse into stillness.

Time passes; a club server approaches unbidden. Removes the used cup. Another server comes to place a clean cup at her right hand. She hopes her brother will not object to the extra expense. Tea given to a stranger who appeared to be tired.

"I couldn't eat anything – I am glad you did not order cakes – did you?" As he questions her, the brother gestures for the wicker armchair to be positioned where he wants it. He also faces the view. As he stretches out, the wickerwork gives out those characteristic, resentful creaks. Too late, she realises how much she had liked the old chairs at home that had received a body in silence.

Talk is of servants of which she has no experience whatever. How at Baroche, Surat and Bombay, a Hindu will not even remove from the table a dish of beef. Nor one defiled by beef. For a Mohammedan, pork an anathema. For a Parsee, hare or rabbit. A Parsee cannot snuff a candle for fear of extinguishing his deity. One to light, one to lay, one to bring, one to serve: interesting enough to hear, but not bound to be pleasing to a person in straitened circumstances. They cannot know she will be interested: in all the differences it might be possible to learn to distinguish in the crowd. Were they trying to demonstrate knowledgeability? Or trying to exclude her?

The names of servants experienced roll off their tongues with a glibness she will surely never match. If she had known servants, she would be vague conversing. Remembering instead personalities, incidents. Hearing inwardly the voice. Seeing characteristic movements. Drift on recalling. And be much more boring to listeners than these women, all theatre. Ask them about their present servants on next meeting

and they might say they know nothing of them. With look of a what-did-she-take-them-for?

The blank end of a three-and-a-half-storey building is plastered. Newly plastered in three watermarked bands. And painted with pink flowers at intervals as if the surface were cretonne.

Like the free use of flowers, languages scatter the land… thirteen separate languages in the District. Perhaps two hundred distinct languages on the continent of India.

Ruling such intricacy, the peremptory… bring me instantly my whisky… make clean yourself. Rudeness to servants who keep clothes so white. In such a dust. Bearers with seemingly no time in which to look after themselves, for picking up after others who were sometimes drunk-sick in their beds.

For the memsahib with any money at all, nothing to do to fill compatriot's days; bearer, chaukhidar, bheestie, mali, dhobi, sweeper. A subdivision of service to cope with every mortal aspect of existence. It is laughingly said only the family cat has no boy.

Morose face of the sweepers – nothing could be so expressive of expected, and craved, invisibility. Anonymous, creeping figures. Slipping in and out. Replacing receptacles.

Fresh receptacles might be ignored. That the bathroom has been entered, hidden under the thought that it is to be expected. But not ignorable, the eerie regular replacing in regimented order the toothbrush, the comb, hairpins. Accomplished almost as one's back is turned.

As far as local festivals indicate, for the Indian to be ruled by white-faced men who use pocket handkerchiefs is no more surprising than if the sahibs had been blue faced or elephant nosed. The flick, wipe, sometimes snort, and tuck away of the handkerchief; symbolic gestures, overpowering, casting spells. It can be supposed that the soldiery were first seen as rhythmically stepping dancers. Red to the waist, thereafter tightly encased, even to the feet. The suffocatingly thick clothing, some kind of ritual binding, the wherewithal to achieve mastery.

Occasional glimpses of the nine nights of plays. And nine days of fun-fairs and red powder throwing, leaves the girl with an impression of their acting out some past ruling cast.

The final procession will bring together the actors. Fireworks will be lit. Great cane and paper statues consumed as a popular spectacle. To be observed from the customary vantage point of the racecourse bund. Picnic suppers the order of the day.

The embankment that carries the canal services road, bounds the better part of the race course. The centre of the course is, by tradition, the polo ground.

In the club enclosure, parties, some already quite jovial, sit on travelling rugs spread on the earth. Or in Bengal camp chairs; invariably awry due to the slope.

On the terraces, club servers minister to a buffet supper. For picnic supper parties, chilled claret betokens membership of the Nantucket Ice Club.

Now, it is past sunset. The embankment crowd's sound quietens for a spell. As if for the first time wishing to notice the feverish activities of the Indians on the polo ground. As if by common instinct, expecting the start of the spectacle.

Are events about to begin? Statue erectors work with undiminished zeal at their flimsy structures; gesticulating amid milling onlookers. The crowd's noise surely has a higher pitch.

The girl has refused the elevation of a Bengal camper, and sits instead on a camel blanket. Each body within her limited field of vision is repeatedly examined for any suggestion by the form that it might be he. That she cannot see the man with any other party affords her some relief. She knows with self-reproach that, three weeks

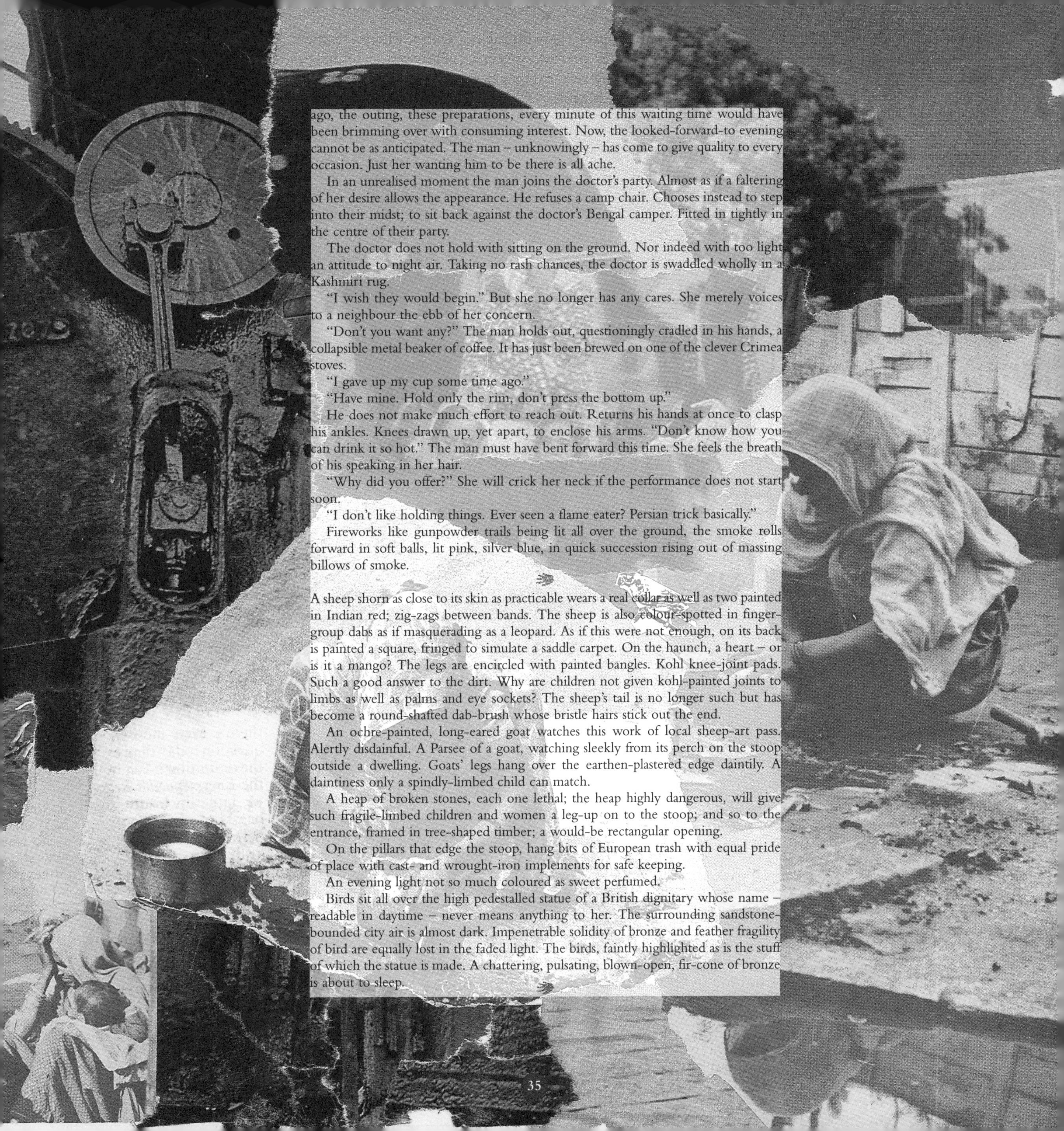

ago, the outing, these preparations, every minute of this waiting time would have been brimming over with consuming interest. Now, the looked-forward-to evening cannot be as anticipated. The man – unknowingly – has come to give quality to every occasion. Just her wanting him to be there is all ache.

In an unrealised moment the man joins the doctor's party. Almost as if a faltering of her desire allows the appearance. He refuses a camp chair. Chooses instead to step into their midst; to sit back against the doctor's Bengal camper. Fitted in tightly in the centre of their party.

The doctor does not hold with sitting on the ground. Nor indeed with too light an attitude to night air. Taking no rash chances, the doctor is swaddled wholly in a Kashmiri rug.

"I wish they would begin." But she no longer has any cares. She merely voices to a neighbour the ebb of her concern.

"Don't you want any?" The man holds out, questioningly cradled in his hands, a collapsible metal beaker of coffee. It has just been brewed on one of the clever Crimea stoves.

"I gave up my cup some time ago."

"Have mine. Hold only the rim, don't press the bottom up."

He does not make much effort to reach out. Returns his hands at once to clasp his ankles. Knees drawn up, yet apart, to enclose his arms. "Don't know how you can drink it so hot." The man must have bent forward this time. She feels the breath of his speaking in her hair.

"Why did you offer?" She will crick her neck if the performance does not start soon.

"I don't like holding things. Ever seen a flame eater? Persian trick basically."

Fireworks like gunpowder trails being lit all over the ground, the smoke rolls forward in soft balls, lit pink, silver blue, in quick succession rising out of massing billows of smoke.

A sheep shorn as close to its skin as practicable wears a real collar as well as two painted in Indian red; zig-zags between bands. The sheep is also colour-spotted in finger-group dabs as if masquerading as a leopard. As if this were not enough, on its back is painted a square, fringed to simulate a saddle carpet. On the haunch, a heart – or is it a mango? The legs are encircled with painted bangles. Kohl knee-joint pads. Such a good answer to the dirt. Why are children not given kohl-painted joints to limbs as well as palms and eye sockets? The sheep's tail is no longer such but has become a round-shafted dab-brush whose bristle hairs stick out the end.

An ochre-painted, long-eared goat watches this work of local sheep-art pass. Alertly disdainful. A Parsee of a goat, watching sleekly from its perch on the stoop outside a dwelling. Goats' legs hang over the earthen-plastered edge daintily. A daintiness only a spindly-limbed child can match.

A heap of broken stones, each one lethal; the heap highly dangerous, will give such fragile-limbed children and women a leg-up on to the stoop; and so to the entrance, framed in tree-shaped timber; a would-be rectangular opening.

On the pillars that edge the stoop, hang bits of European trash with equal pride of place with cast- and wrought-iron implements for safe keeping.

An evening light not so much coloured as sweet perfumed.

Birds sit all over the high pedestalled statue of a British dignitary whose name – readable in daytime – never means anything to her. The surrounding sandstone-bounded city air is almost dark. Impenetrable solidity of bronze and feather fragility of bird are equally lost in the faded light. The birds, faintly highlighted as is the stuff of which the statue is made. A chattering, pulsating, blown-open, fir-cone of bronze is about to sleep.

Chapter 6

BUILT PLACE

The wash-room, entombed in the middle of the semi-basement, presented a haven of coolness. It is a relief to be able to descend into this at noon, out of the baking city. The broad flight of white marble steps leads down as if into a tank. The whole of the white-tiled cavern shines with yellow light from the three lamps on brackets above the line of basins. Lamps that send their winking, glinting light to the farthest cushion tile. The wash-room servant in strange white hat lays out for each visitor a tiny tablet of over-scented soap, a towel, and a cloth. Addressed in what sounds to the younger man like Pushtn, the peon grins. The older man washes his face. Then holds to the back of his neck an instant the wrung-out hot cloth handed to him by the peon. Closes his eyes and holds the cloth across his nose. Dries himself thoroughly, almost ritually.

In the dining room, a huge metallic mirror comes down to dado level beside their table. Reflected in it are the green-grey of the blinds drawn over the windows to the street. The decor, of a period with the building, is now of faded sand colours, picked out in tobacco colours. Doric columns compartment the room to make a tripartite space in the style of the portico to the club.

The open area that extends between tables dominates. Each table floats on its own. Copious white linen covers each table, falling in heavy folds to touch the floor. The majority of the tables hold placements for a single chair. Those at table are ministered to by three servants in such a manner as to suggest these are brethren of those attendants on Venus rising from the sea: something to do with gentle gesturing service enacted against the folds of napery.

One group of men are looking at the building, discussing; what do they say about it? How much think or know of the history of the continent? Two men, perilously close behind this chattering group, squat, talking. On the ground at their side a tin battered beyond belief, but one hand – all extended fingers – moves it lovingly, expressively. Others, milling in and out, move forward to the terrace to look at the non-existent view, calling to each other to come and take great interest in the mud and scrub visible before an obscuring haze.

Always as his carriage approaches the English Church, the last of the marching soldiers pass heavily before him. Crunch to a less distant standstill, and to shouted orders, pass in at the front entrance. Heard at the transept entrance, the determined thuds as the Snider rifles are placed in the sockets of the hymn book rack. These sounds strike the most real notes of the routine of church yet are heard remotely, through doors held open. A meagre strip of coconut matting leads to the accused-style pew.

Those noises are the reason why he dislikes attending church: why he has reduced his attendance to the minimum the appointment demands. To be so firmly, resoundingly set apart from contamination by lowly people makes his flesh creep; too much like hearing the lid knocked down. He knows that he alone thinks in terms of deprivation; but he cannot shake off the sense that churches are for solemnising states of the person.

The thin steely spikes of the bayonets glint in the mauve and yellow light, from the new stained-glass memorial windows. On this Sunday the coloured reflections constantly flicker, drawing his attention; nauseating to watch. Occurring outside,

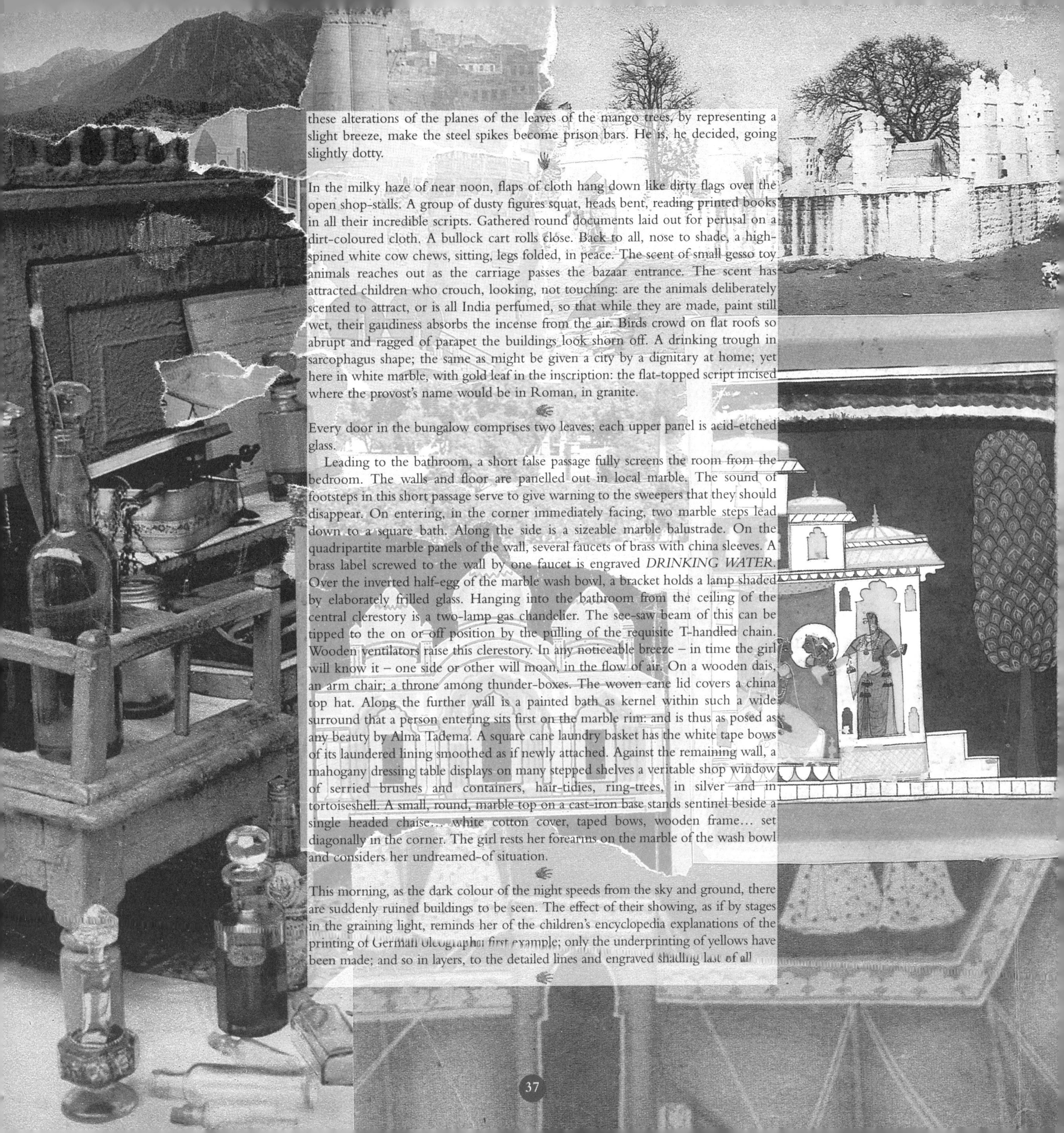

these alterations of the planes of the leaves of the mango trees, by representing a slight breeze, make the steel spikes become prison bars. He is, he decided, going slightly dotty.

In the milky haze of near noon, flaps of cloth hang down like dirty flags over the open shop-stalls. A group of dusty figures squat, heads bent, reading printed books in all their incredible scripts. Gathered round documents laid out for perusal on a dirt-coloured cloth. A bullock cart rolls close. Back to all, nose to shade, a high-spined white cow chews, sitting, legs folded, in peace. The scent of small gesso toy animals reaches out as the carriage passes the bazaar entrance. The scent has attracted children who crouch, looking, not touching: are the animals deliberately scented to attract, or is all India perfumed, so that while they are made, paint still wet, their gaudiness absorbs the incense from the air. Birds crowd on flat roofs so abrupt and ragged of parapet the buildings look shorn off. A drinking trough in sarcophagus shape; the same as might be given a city by a dignitary at home; yet here in white marble, with gold leaf in the inscription: the flat-topped script incised where the provost's name would be in Roman, in granite.

Every door in the bungalow comprises two leaves; each upper panel is acid-etched glass.

Leading to the bathroom, a short false passage fully screens the room from the bedroom. The walls and floor are panelled out in local marble. The sound of footsteps in this short passage serve to give warning to the sweepers that they should disappear. On entering, in the corner immediately facing, two marble steps lead down to a square bath. Along the side is a sizeable marble balustrade. On the quadripartite marble panels of the wall, several faucets of brass with china sleeves. A brass label screwed to the wall by one faucet is engraved *DRINKING WATER*. Over the inverted half-egg of the marble wash bowl, a bracket holds a lamp shaded by elaborately frilled glass. Hanging into the bathroom from the ceiling of the central clerestory is a two-lamp gas chandelier. The see-saw beam of this can be tipped to the on or off position by the pulling of the requisite T-handled chain. Wooden ventilators raise this clerestory. In any noticeable breeze – in time the girl will know it – one side or other will moan, in the flow of air. On a wooden dais, an arm chair; a throne among thunder-boxes. The woven cane lid covers a china top hat. Along the further wall is a painted bath as kernel within such a wide surround that a person entering sits first on the marble rim: and is thus as posed as any beauty by Alma Tadema. A square cane laundry basket has the white tape bows of its laundered lining smoothed as if newly attached. Against the remaining wall, a mahogany dressing table displays on many stepped shelves a veritable shop window of serried brushes and containers, hair-tidies, ring-trees, in silver and in tortoiseshell. A small, round, marble top on a cast-iron base stands sentinel beside a single headed chaise… white cotton cover, taped bows, wooden frame… set diagonally in the corner. The girl rests her forearms on the marble of the wash bowl and considers her undreamed-of situation.

This morning, as the dark colour of the night speeds from the sky and ground, there are suddenly ruined buildings to be seen. The effect of their showing, as if by stages in the graining light, reminds her of the children's encyclopedia explanations of the printing of German oleographs: first example; only the underprinting of yellows have been made; and so in layers, to the detailed lines and engraved shading last of all

Chapter 7

REPEATED INCIDENT – PLACE REPEATED

By the sound, the man comes out of the room and commences walking. The girl turns to the sound of the approaching brisk footfalls.

The man stops abruptly, as if in afterthought; his stopping unpremeditated. He stands, separated from the girl by the symbol of her authority, the spindly-legged desk. Already the heat smites through his clothes. On this side of the building, the afternoon sun penetrates and further reflects inwards in great bright planes. This visit to the plains, the ability to adapt seems lost. At his temples, pulse beats are visible. The man is aware of this, the pulsating blood is almost squeaking in his ears. "I am about to depart." Even to himself he sounds breathless.

A rather forced smile. "I hope you have a pleasant journey, Colonel Nicholson. Maybe… if you take another furlough here…" The girl gestures towards the closed door as if he would understand some meaning.

But he does not understand, forgets what it is, if anything, he is about to say.

The girl laughs at the silence: and could more easily have wept. She wishes the bell in the civil cantonment would ring riot… a dramatic role; in the performance of which she could so have acquitted herself that he could have admired her.

As it is, he stands perplexed. He straightens his shoulders, and in so doing takes a deeper breath.

He is going now – she thinks, brightly gritting her teeth in a smile. Although she watches intently, his steady gaze betrays nothing… whether he dislikes her… does not consider her. Nothing in his manner cheers her spirits. Except that he remains where he is.

The man shifts his stance. Looks at the foot now thrust to one side. Stares towards the entrance. Does the smoothness of his face express a vacant mind?… disdain? The gaze is outwards, downwards, at the beaten earth of the compound.

Again the man shifts his position.

The girl takes a step aside from the desk.

Gradually Nicholson turns so as to continue facing her. By stages turns through a semi-circle. Nothing the while centres in his head to say. He blinks at the glare he now has to face. "One picks up again with acquaintances where one left off," he hears himself remark, "… years after." Even repeated to himself in thought this does nothing to explain itself, or to reassure him. Has he seen a look of anxiety on the girl's face?

Raising her hand she brushes something off her neck… but she is looking along the empty verandah.

Nicholson comes to a stop in front of the stand-up-and-beg desk in the dispensing hall. By the contrast to its surroundings, the girl's sprigged muslin dress indicates a suitable spot to pause.

A passing light-headedness… the temperature standing as it did at 108 degrees in the shade; steadily 15 degrees above that in the hills to which he is accustomed.

However, since he has stopped – to make his absent-mindedness less dotty – something has to be said.

"A long-standing invitation…" The man hesitates, seeing the drawn-down bow of her mouth. Why does this information concern her? "A couple of indigo planters, in whose District I once did a term."

A move aside a pace, following a suggestion that a breath of air existed, she took as his desire to leave.

The last thing he wants is to step into the heat coming off the earth outside. Therefore, resolutely, he stops… stares out at the unwelcoming haze and the micaceous brightness.

Occasionally the abrupt, dipping flight of some bird.

A whole host of green pigeons in a small tree. How dead the landscape would be if Hindus permitted killing!

A group of men hesitatingly comes up the steps, salaams thrice. Wide-pressed, dusty sandals deposited – as is the custom – on the lower verandah steps.

"I thought you would be here on Wednesday evening. The Regimental Band plays in Lawrence Park before dinner."

He looks at her in amazement; even had he known, it would not have entered his head before accepting Sunter's invitation. Exasperated, the man exhales a long breath. Sees his boots are coated with pollen dust and is depressed by the sight. He stares all round, hands clasped behind his back. He braces his shoulders, and breathes deeply again. Looks at his boots' rounded toes; then at the dust raised by the approach of an entire family. The group salaam thrice; and are ordered in abruptly by the chuprassi.

"Out here you will find you can remember even slight acquaintances, six or seven years after first meeting."

As if in response, she laughs. In short gasps at a rather high pitch. On his looking at her, she exclaims, "But you will be ancient and walking with a stick by then, Mr Nicholson."

What had this supposition got to do with when he's last seen Sunter? "I don't follow," he tilts his head. Then as immediately, regrets this as too like a gesture an aging man might make. He blinks, and turns back to staring at the earth outside.

The sun's light is now perhaps more brazen. The dust haze that is high on the horizon will in a minute mercifully veil the sun; turn the sun now lost in brightness to a fiery disc; magnifying its size.

He cannot connect together the broken pieces of conversation. "Well," he says lamely, "You never know."

"Do you wish the letters returned?"

"No."... To be suddenly faced with such a decision....

"Why should I want them?"

Without saying anything he regards her reproachfully.

She laughs, delighted that this man has taken possession of the letters. "You surely don't expect me to keep letters in a box, tied with ribbon, and have myself reminded of all this sadness at odd times of opening drawers in my room?"

The man finds to his surprise that he resents her lack of sentiment. Someone should care. Did a youth live nineteen years, and only those at home, thousands of miles away, feel any concern at his passing? In mortification, the man leans forward against the verandah post. The weight of his body leaning on the post is taken in the forward hollow of his shoulders. His hands are clasped behind his back.

"I am sorry about Mr Nicholas."

"I imagine you are." Why should it make him so angry, this nuisance of Nicholas's passing?

"But I would prefer not to live mutilated."

He cannot think of replying. He is saddled with uncovering the full extent, the portent, of her indiscretion in writing to a youth in a ward in the military hospital. Had his brother, or another, opened the letters? Had his brother read them? Taken out of their envelopes they had been. But by whom? By how many?

"Do you write to all your acquaintances?"

"The boy was wounded."

Until he sees the expression of the girl as she sights him, convention has suggested the return of the letters as proper duty. Consideration of any action other than return had not entered his mind. Now he is smitten with a realisation of the disturbing nature of his errand.

He goes forward more slowly. "Miss Urquhart – I – have some letters I believe are yours." As he speaks, watching her face, his hand reaches into his inner breast pocket. Having extracted the three identical envelopes, he holds them out. "The thought did occur to me you might be concerned as to where they had got to."

How shocking to have handed back envelopes once posted. As if the proper course of events has gone wrong: which of course is the case. Merely touching the corner of one with her finger, the girl takes a step backwards, as if repelled at the sight.

"You should not have bothered."

"They were among his effects. I am nervous lest they be read by others."

"I wrote nothing that could not be."

"Personal letters are not the sort of thing for other eyes."

"Your brother lay hurt; I could not visit the military hospital."

To come all this way. To this godforsaken, flat, dust-covered, fly-blown fate-chosen place. That this strange subaltern should be stationed here. To leave as legacy this distasteful special delivery.

"... if only because of the close association you had with my brother."

"No such thing."

"Well that is perhaps a debatable matter... contrary to evidence."

"I'm sorry you think I switched my friendship in too unseemly a fashion from one brother to the next," she said hotly.

"Your words, Miss Urquhart, not mine."

"But that is what you are meaning, is it not Mr Nicholson?"

The grey eyes look at her.

The habitual backward tilt of the head, accentuating any intended arrogance.

"That's unkind, unjust: indeed it is, Mr Nicholson. What person claimed I had an association with your brother, whatever is meant by that?"

"Not only from what has been said, Miss Urquhart." The man reaches into his inner pocket. "Do you wish the letters back?"

"No," she steps back. "No, they are no good now. Either they cheered, or they did not. It is of no consequence now."

The man still holds the packet out.

"You should have destroyed them," she adds sadly.

The man turns the packet slowly in his hand without taking his eyes from the face. A face that is melancholy; yet not so. That is not the usual pale English face. A face that is full; that does not indicate how it might show character due to age. The eyebrows, in concern, pucker two deep wrinkles above the nose, the mouth is a down-turned bow of slightly ridiculous seriousness.

"You do not consider it strangely cruel to destroy for yourself all evidence that he existed?"

In truth, at this moment, the girl is far more unnerved by this older man speaking with her than by any realisation of loss. Only later that evening, in recalling this scene, will memory burst out in angry tears. Tears as abruptly stopped by the relieving thought of how fortunate no such outburst came that afternoon: how extraordinarily unfeeling of the man not to have foreseen this possibility.

"All his belongings are dispersed. A distasteful duty, to wipe away all evidence of a man's existence. Yet this I've done. All within the space of a few days of my having arrived here. These I cannot send home." Indeed, what balks Nicholson is unconnected with the letters. The grave omission is that of a young man; nothing accomplished, nothing contributed to the continent, save the stone in the graveyard.

"What have you done with his things? Did you keep anything?"

"I'm sorry, did you want something?"

"No."

"I should have asked… "

The girl shakes her head.

"… but, to begin with, the letters were all I had to go on."

Inexplicably, and quite inappropriately, the downturned bow of her tightened mouth makes him want to laugh.

The girl looks away from his disparaging gaze.

"Where did everything go?" The problems left by the dead interest her… one needed to know one's role… the church held the service, friends gave condolences…

"If there was something particular…?"

"I am only curious."

That she has seen with such prim surprise the amused expression of his of a moment ago, further erodes his patience. "His fellow subalterns auctioned his kit, as you must know is the custom: his Colonel collected the few personal belongings. I gave some of these to his servants and had the remainder boxed home."

"You make yourself sound callous."

"I have done everything required of me. I had seen him exactly once, briefly, since he came out. Myself, a mere seventeen when I left home. What brotherly feeling am I expected to conjure up for a toddler who next appears as a pale, moustached – which I cannot abide at the best of times – young subaltern, claiming me as a relative; mouthing places and people I know only through letters from home, or of which I hold the vaguest of memories." The speaker strikes the nearest post of the verandah with his fist. Then grasps the turned, blue-painted post. Arm stretched out at a high angle. Since his arrival in the plains there has never occurred the right moment, or the right audience.

The sense of frustration compounded by the pomposity of his compatriots, the heat. "I have spent the better part of a week suffering pointless commiseration. Now I am finished." His hand, removed from the post, cleaves the air. "Finished. I refuse to be put in the position of having to pick up after him… the Bible is full of that sort of filthy deal." He turns his back on the girl. Walking away a few steps before grasping the verandah rail firmly in both hands and leaning angrily, heavily on straight arms.

"As that is undeserved, it is extremely unkind, and I thought you a kind person."

"You must not judge from so slight an acquaintance." The issue is both childish, and of some principle. He turns his head and looks at the girl.

The two people glower at each other.

Nicholson looks away, cross with himself for bringing up the business of the letters. He should have dealt with them quietly.

"Surely it has nothing to do with Mr Nicholas that I thought we might become friends: unless you want to quibble."

Perhaps, he thinks, he has become too attuned to this country of his adoption, has adopted a too studied habit of sliding without travail through well-organised days. Yet the hermetically sealed life in the frontier service is exactly to his tastes. The tantalising daily routine of preparedly awaiting the inevitable pot-shots from among the rocks.

By contrast, the present elusive arrangement practically unhinges his mind. Around him swirls a situation not opted for. That cannot be manipulated by greater powers of detachment. Beyond the control of firmness of voice. No clarity of instruction can bring events back to a well-rehearsed course.

"I must assume you feel a real antipathy to women, Mr Nicholson."

Shocked, Nicholson averts his gaze to a spidery tamarind whose foliage gleams evilly behind her back, hinting at unabated heat.

"I am not sure I can accept that remark, Miss Urquhart," he says smartly. "I simply see this as an issue as old as the Bible itself. The Commandments are quite clear."

"I was not a… a belonging – like a lamb or a goat-hair tent."

Nicholson descends two steps to hide an involuntary smile. However, his back is not so long turned to her as to be rude.

"You are quite unfair, Mr Nicholson, I purposely avoided any situation that could give rise…" The girl, prompted by the action of the man, descends three steps. There, she turns round to face him.

As the girl descends the steps, the man has to pivot on his heel. He does so, on the step he stands on, with some difficulty.

"Your brother was far too young for me."

Nicholson draws in his breath sharply. Young! The man looks over the girl's head at the heat fairly smoking off the ground. This rising, heated air, wobbling leaves. Setting the foliage glittering. Upturned yellow soil, falling on a heap, turned out from some ditch being dug in the grounds, turns paler almost as he watches each basket-load tumble to a trickle of earth. Another sanitary improvement. They were everywhere. The rage since the Crimean experience. Recalled by the pattern of his thoughts to some sense of duty, the man makes a suggestion. "Well, perhaps you want to have the letters and burn them."

"No."

To bicker so made him lose grip on the purpose of his visit.

"I think the sense of them is more friendly than you realise. I had to read them to him."

"They were written to try and comfort."

"I have never received letters like them." Immediately Nicholson regrets this opening of his flank.

More upsetting to him, the girl seizes on quite another aspect. "I am not surprised, if you are as ill-disposed to everyone who tries to be your friend. After your brief visit, what kind word did you write?"

"I probably would have got round to writing, offering a chance to spend sick leave. Nothing else I could have done."

"And then he died."

"They telegraphed me he'd died." Nicholson descends the steps, passing the girl standing on her step, still facing inwards to the dispensary.

Chapter 8

HORSEDRAWN

Some memory smites; and as quickly flees. The horses around their carriage clip-clop on their hasty way. Wheels whirl their spokes beside their trundling wheels. Two opposing streams of wildly assorted wheeled vehicles. Along either side the crowd spills on to the thoroughfare, in places almost blocking the path of the outermost stream of carts; to be constantly shouted at, whip cracked.

With an effort of will the girl brushes away these crowding presences, the cacophony of noises. She wishes to search for that memory which touched a chord a moment or so ago; even to the extent of turning and looking back along the route they have come. Whatever sight acted as prompt is not seen again. Still she wants to catch that fly of memory on a carefully reconstructed cobweb of her thoughts. What had it been?

All four passengers are rocked, side to side, by the motions of the tonga. Each head is turned to their nearest side. The girl, who since her arrival has felt squeamish, is most conscious of the smells of rich food. Meals are cooked by the wayside, with passers-by served from great flat pans in which the mess is ready and glistening; dark red, rich ochre. Food is eaten in situ, as the crowds surge through each other. All along the way, greasy aromas.

Trotted and rolled past all this, their tonga draws abreast of a worn-out beast… the girl turns her head away from the sores visible under the stuffing-spewing collar; only to see a man sleeping where he might easily lose a limb… head is turned aside once more. She tries to think what the dust smells most like: scented somehow, a scent made into things; or is it just less than scented? Everyone in the quarter seems to be going to the opposite end of town. What stirs them so?

Almost immediately they are in the hired conveyance, the mad start made by the driver makes them realise they have given themselves into the clutches of some too willing, untamed animal. The man cannot be restrained. His animal has worn itself out without bringing about the slightest change in its master's insistence on mad speed. Is he about to carry them off through the densest part of the native quarter? Only by concentrating on retaining their composure of face, or evenly returning the startled and concerned gaze of the few compatriots they race past, can fear be rationally fought. Her brother repeats his admonitions at regular intervals. Now firmly. Now cajolingly. Then bye the bye. Again, sternly calling for less speed.

Noises of horse harness moving fast… tongas in haste? Slow harness chimes… a carriage closed against the heat? These noises jangle out beyond the metal-pan sound of that bird. Located by a belt of dust raised by this incessant passage, dust contained within the trees bordering the wide road.

Beside her, in an aggressive movement suddenly made, Nicholson stands off the post.

A group of orderlies passes the verandah end. The girl watches the three figures in white cloth, brown tails of head cloths, until they disappear from sight.

Birds are making such a racket in the trees. And, somewhere in the native quarter, extraordinary music. As if beans are tossed on a metal tray, accompanied by bells on a fiddle bow. Now and then, joining in, a sound like teacups shuffled on the tray. The background rhythm is as if whipping up a mixture whose recipe is called out occasionally in matter-of-fact tonal voice.

The girl considers she had better speak further as to the reasons for her taking the liberty of writing. "A skirmish they called it. Such a terrible consequence probably only happened to half a dozen soldiers in the Siege of Delhi."

"More common than you think. One ball took off the arms of two men, side by side in a window, at Delhi."

"Did you know either?"

"No. I had just started in the Punjab at the time." The man now remembers he had started by making a point: she has lost him completely… "So, I am to destroy them?" He looks at her; it does not now matter if she had hoped his brother would marry her: no concern of his. He sighs and steps down to stand on a lower step.

"Friendship is more than talk." However, this is not said belligerently.

Already the situation has reached beyond her experience. The girl feels helpless. She watches him as he leans against the post; leans back against the painted post that brackets the verandah roof out as hood to the three steps. Lounges, so that his feet are crossed at the ankle. Legs straight, bracing his leaning on the post. The angled pose shows off the cut of his suit remarkably well.

Heat, direct from the lowering sun, strikes through his coat, touches his shoulder. Palpably strikes, on the side of the head, as far as just above the lobe of his ear. Not unpleasant. No one would take this amiss as long as it does not strike the spine: he has noticed all the children in the cantonment wear spine pads. At least he had not been a child out here; such an encumbrance must be uncomfortable.

The space in the rafters of the verandah roof is busy with flies.

Her hand is repeatedly raised, as if to ward a stray insect off her hair.

Beyond the handrail is an empty scene. Stretching his neck out of his collar with a movement of chin, the man turns his head in so doing. A figure on a charpoy is under the tree. The body sleeps on its side, hands praying under its head.

"Well, I must not keep you."

The man looks at the girl and sees a sickening smile.

Goodbye, Mr Nicholson," and as immediately, she turns and, through a haze of her own sadness, makes her way to the place where she knows her desk to be. She cannot shirk work, with a good conscience, any longer.

To cease relaxing, to go out into the heat is not at all what the man wants to do. Crazily he wants to turn and protest at the abrupt dismissal. Seeing no redress without embarrassment, he gives up. Reluctantly the man crosses to the shade the driver of his gharri has been instructed to keep within.

The cavalry is loosely strung out as yet. Syces are walking the horses. The smell of horse mingles with that of watered dust. At the Mall side is drawn up the

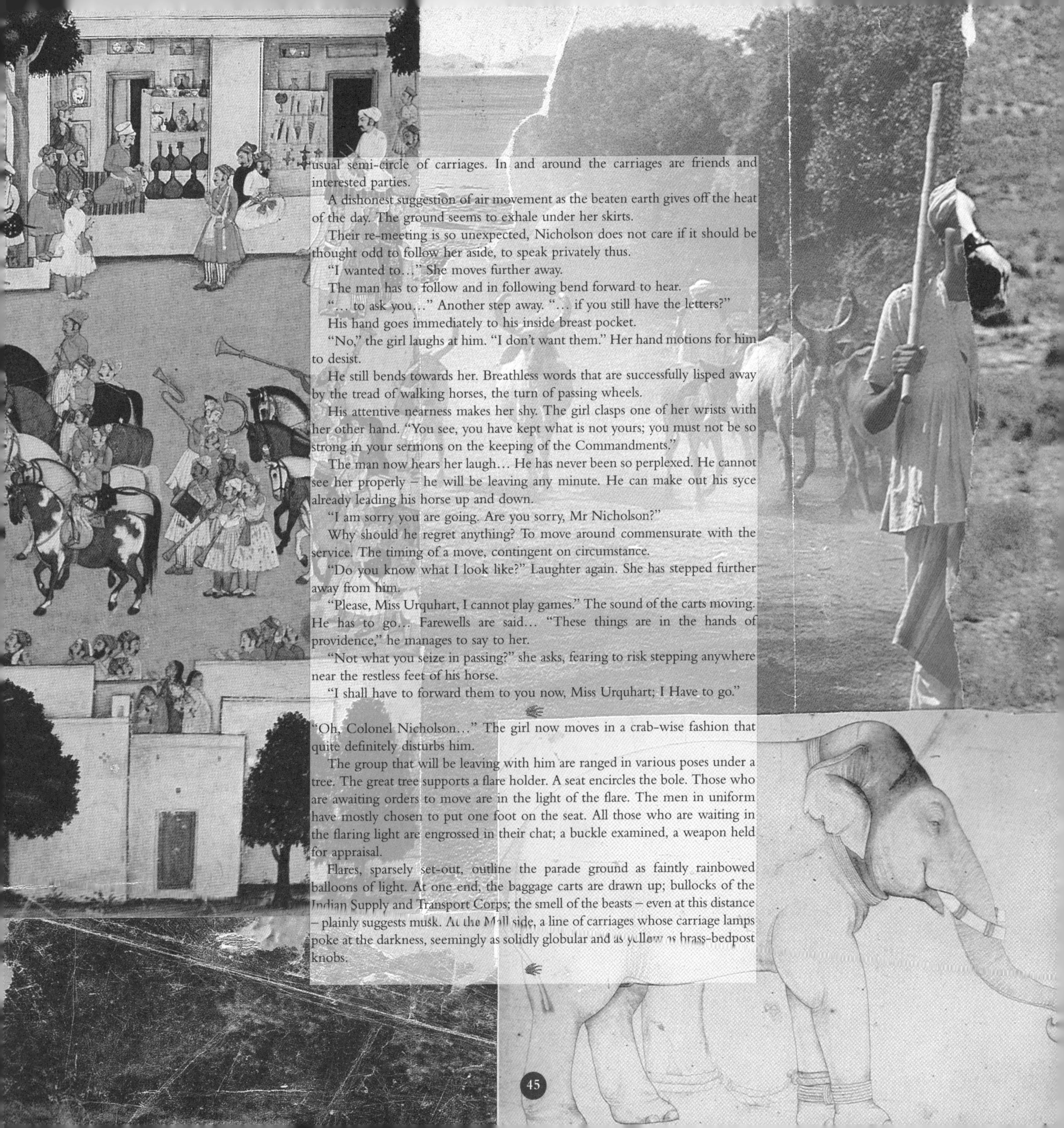

usual semi-circle of carriages. In and around the carriages are friends and interested parties.

A dishonest suggestion of air movement as the beaten earth gives off the heat of the day. The ground seems to exhale under her skirts.

Their re-meeting is so unexpected, Nicholson does not care if it should be thought odd to follow her aside, to speak privately thus.

"I wanted to…" She moves further away.

The man has to follow and in following bend forward to hear.

"… to ask you…" Another step away. "… if you still have the letters?"

His hand goes immediately to his inside breast pocket.

"No," the girl laughs at him. "I don't want them." Her hand motions for him to desist.

He still bends towards her. Breathless words that are successfully lisped away by the tread of walking horses, the turn of passing wheels.

His attentive nearness makes her shy. The girl clasps one of her wrists with her other hand. "You see, you have kept what is not yours; you must not be so strong in your sermons on the keeping of the Commandments."

The man now hears her laugh… He has never been so perplexed. He cannot see her properly – he will be leaving any minute. He can make out his syce already leading his horse up and down.

"I am sorry you are going. Are you sorry, Mr Nicholson?"

Why should he regret anything? To move around commensurate with the service. The timing of a move, contingent on circumstance.

"Do you know what I look like?" Laughter again. She has stepped further away from him.

"Please, Miss Urquhart, I cannot play games." The sound of the carts moving. He has to go… Farewells are said… "These things are in the hands of providence," he manages to say to her.

"Not what you seize in passing?" she asks, fearing to risk stepping anywhere near the restless feet of his horse.

"I shall have to forward them to you now, Miss Urquhart; I Have to go."

"Oh, Colonel Nicholson…" The girl now moves in a crab-wise fashion that quite definitely disturbs him.

The group that will be leaving with him are ranged in various poses under a tree. The great tree supports a flare holder. A seat encircles the bole. Those who are awaiting orders to move are in the light of the flare. The men in uniform have mostly chosen to put one foot on the seat. All those who are waiting in the flaring light are engrossed in their chat; a buckle examined, a weapon held for appraisal.

Flares, sparsely set-out, outline the parade ground as faintly rainbowed balloons of light. At one end, the baggage carts are drawn up; bullocks of the Indian Supply and Transport Corps; the smell of the beasts – even at this distance – plainly suggests musk. At the Mall side, a line of carriages whose carriage lamps poke at the darkness, seemingly as solidly globular and as yellow as brass-bedpost knobs.

Chapter 9

SCENTS AND SMELLS

To begin with, dawn reveals a landscape of earth worn to powder… empty… everything absolutely still. Then come treading animals. These are led out into the landscape by figures of no thickness. Clad in gently flapping linens. Woven of fibres the colour of the dried-out earth-dust. Each dry body – the self-same colour as the earth – is slipped in, and in its stick thinness is sticking out of these loose wrappings. The knees of the men rise, the foot is parallel with the earth. A long stride branch is left behind some nether folds. The end of a cloth hangs down, which ever knee is raised.

Flesh so hungers after water that the poor cube of a plastered earthen room, set aside for the lota and its use, becomes very pleasurably entered. At first, she had felt surprise at such primitive conditions. This has become her favourite room, although more like a bothy in its earthen floor and the whitened walls… the fresh drips of whitewash overlie old, and older, speckle fringing to the hand-smoothed undulations of the floor's edge. By day the room cube starchily suggests freshness. All internal surfaces are equally white. From the high and tiny under-eaves window comes an ample flood of light, abetted betimes by a narrow cross-floor beam of light piercing through the heavily grated outlet. Light, softer and more velvety by evening, illuminates a room then humming full with biting insects either entered via the grating, or penetrated through the door. In early evening's dusk, the little smoking oil lamp is best lit to keep such pests in half hiding. The all too resonant brass pot has an upturned earthenware bowl for its cover, to keep out crawling insects. This pot, shined daily in ash, contains the entire water supply for all functions. All spent water finds its way out to the sweeper across the floor. The floor's surface dips away from all the walls and traverses water towards the grated outlet. She never knows if the sweeper sits against the wall. Can sounds be heard the other side of the little grated hole that shines a blob of morning sunshine on the beaten earth? Maybe sound does not matter. The sweeper can sometimes be seen from the orchard, tending his fire for cleansing pots… with his hoe grooving the hard earth to lead irrigation to the young trees. Maybe such a servant becomes intuitive of the room's use. She hopes she is no trouble: some old India hands never pass a day without a mild attack of Bengal-belly.

He looks around in desperation. A step to the right would precipitate him into the pond. A move backwards might bring him too near the circle of light cast by the large porch lantern. Any movement might attract the gaze of the waiting carriage servants.

The stationary, hired conveyances screen the lit porch from the garden only fairly effectively. The night is inky black. Oppressively, smotheringly dark. He is more than relieved when she steps aside.

She leaves space for him beside the plant-pots getting their water at the pond edge. She proceeds to promenade along the pond's length.

Intently he watches the carriage servants for fear the movement of her dress of light stuff will place them under their attention.

In fresh fright he remem-

bers that in such a garden there will be a Watchman… To be shot at by an old rusty musket… To be bitten… The man looks round wildly. But sees only the sheen of the surface leaves of the shrubs gleaming with moisture. An innocuous wall of foliage. A flipping, tapping, dripping leafy surface, an arm's reach away, only obscurely distinguishable from the humid night.

The girl passes round the pond. Returns to where he stands. The side of her head hits his shoulder quite hard, enough almost to knock him on his back. As he steps backwards to regain his balance he instinctively grips both her arms in case there should prove to be only pond behind him.

"I don't want you to leave."

Leave! He is thankful not to be in the pond.

Sweating earth. Vegetation stealthily growing. Foliage expanding to abundancy. All join with smells peculiar to a black night. The airy amalgam rises up and wraps his clothes tighter to his body at the very instant he steps off the lighted verandah. Busy creaks and croaks accompany his walk across the lawn. Droplets sound in the high trees.

"Come and hear the frogs."

The man nearly jumps out of his skin; but he replies evenly, "I am just leaving."

"I know; I saw you leave. The pond is on your way. This will not take a moment. The frogs will all plop in the water as soon as you get there."

Perhaps a normal suggestion might have been more stupid to comply with… or more easy to refuse… this is a situation the man finds too difficult to oppose.

"Be as quiet as you can." The girl skips across the intervening stretch of gravel drive. She awaits him on the grass beyond.

He can see the light form of her muslin dress against the blackness of the night.

Plop, plop, plop, plop… just as she has described. The man stands in the darkness, looking down quizzically into the rectangular sheen of the water: troubled and marbled by dark reflections. He looks at the sky for sign of a hidden moon. Then – unable to resist his nerves – he glances towards the porte-cochère. And then further back, at the insect feathered verandah lights. But under the lights, the bearers all face inwards; attentive to the scene in the bungalow interior. Carriages awaiting people who will soon be leaving.

"Stand very still, here; I will chase some over." This directive is offered in a most conspiratorial whisper.

He watches as she skirts the pots at the pond's edge. Sure enough, some frogs climb out on to the leaves in front of him. Three try the white marble edge. He can actually make out little forefingers, one aft leg. But they must see him bend low. The limbs disappear. Audible plops scare other dark forms off the floating leaves.

When she stands directly opposite, her lighter form visible across the water of the pond, some frogs climb out onto the white marble edge at either end – not confident – at his straightening… plop, plop, plop. Now, like the frogs, he waits; and certainly he is as nervous. How has he managed to get mixed up in this escapade? At his age?… Why has this girl suddenly spoken to him?

"I like other people's ponds. Here, take this."

She holds out to him something he cannot make out. She grasps his hand and roughly presses a hard stalk in between his fingers. The end of the hard stalk digs down into his palm. The object is so light, the surprise of the hurt so great, he nearly drops whatever it is.

"I like lotus seed pods. Put it in a small vase on your dressing table. A Chinese vase will look best."

Nicholson recognises the girl immediately, although glimpsed through the throng… in the same instant as guessing it might have been her voice describing him. Recognises her surprisedly as the girl who had been promenading the club terrace that morning before breakfast.

Only after a preliminary series of introductions does Cochrane say, "You were asking about a Miss Urquhart. I hadn't realised there existed any liaison serious enough for you to know about."

"I found a letter; that was all."

"Certainly your brother was friendly with the young Scots in the Company office. The girl is a sister of one of those writers."

By the end of the soirée, the girl stands apart, half-listening to the fringe conversation in the room. She stands staring out at the empty verandah. At the night that is beyond the bare boards.

The turn of her head, the droop of the bare shoulders, show to advantage in that graduated margin of light before the open doors. The light splays out of the room, onto the verandah boards. The darkness begins abruptly behind the verandah's balustrade.

Ultimately he withdraws from the crowded room. To leave by the verandah is the quickest exit for a man without his carriage. He hesitates barely a moment before he will pass out of the open doors. The casual, familiar smile he receives irritates; perhaps makes him ask, "Did you not recognise me in there?"

"I thought my not speaking would most likely make you speak to me."

The man's retreating footfalls sound hasty. He has walked across the boards of the verandah with few steps. Steps are heard descending to the lawn. Then silence. But the girl can see he walks briskly away.

As he walks, the man keeps shrugging off remembrance of the smile. What does the girl think she is doing? Behaving like this… ? More trickery after the writing of the letter? Except who would want a one-armed husband, with so many possibilities about with two arms? He resents the furlough wasted, and in a borrowed room in an establishment musty enough to give any sane man the creeps.

The shade of the great trees falls across the open ground… Two men sit on the sparse grass, talking with hand flowering gestures… Three crouch, mending a cart. One lies on his back, knees drawn up… A sleeper's head on a flat basket, its lid open to receive and shade the head. Another, swathed in cloth, might be dead. Two, almost shoulder to shoulder, hardly seem to be together, they look different ways, yet talk; their slight movements suggest a difference of origin, of caste; one wears a turban, the other a lace cap. A dog moves about warily; eyes never on the ground; it searches by nose alone. An old grizzled man rests, head on arms, arms laid across drawn-up knees. Another has a piece of leather luggage, very scuffed and dusty. The official scavengers have their part of the shade and sit minding their pieces of barely recognisable wooden board they use for picking up leaves. Switches are worn to scimitar points. The untouchables sit beside their small cart, fabricated from innumerable, dissimilar, broken, scavenged pieces of wood. Activities of birds never end. Feathers full of energy find blobs of shade in which to hop. Hop, or take little runs, between the resting people.

Utterly derelict, long, dirty verandahs speak of unmarried men throughout the station. A sepulchral atmosphere stresses that decay in this place is taking its toll all the time. A process which may only be arrested by an army of servants chivvied by Scottish willpower.

The surpesh'd servant receives their card on the inevitable brass tray. The dull brass will be tacky to the touch.

As the girl and her brother are ushered in, the men rise from their chairs. They are in a vast, blearily-lit room that overlooks a drop. There is an overpowering odour of cheroot smoke. This is a house without a woman.

Such pitiful wares are for sale. Small squares of aged linen are spread on the ground. Displayed are tiny heaps of herbs, laid out for some checkerboard game to be played with ants making the moves. Successive squatting children offer limes for sale. One boy has five. The girl has a desire to purchase but is afraid she will embarrass both the boy and herself, for her brother might have no change small enough. One lime is quite wizened. She feels she should buy something: otherwise it is as if she is scorning their efforts to earn a livelihood.

But more heart-rending sights are to follow. First, they walk by some vendors of assorted bone combs. These wares are also laid out on old red cotton cloths. Some rags have been prettily sprigged with patterns by Kalamkari dyeing.

Beyond these, neatly set out, are dozens of old pasteboard boxes… ex-Dr Cockle's Pills… some so many times owned their labels barely recognisable. Then follow more things thrown away. A person never expected to see these again. Quinine boxes… blacking bottles, some hardly cleaned or cleanable… Blanco tins. This spoil from kitchen yards comprises quite the largest area of the bazaar.

On retiring that night, he stoops before the mildewed mirror and examines what he can see of himself. His reflection seems distant. A dressing table's width again beyond the canker on the mirror glass. The man reaches out; moves closer. The distance at which the reflection stands is no less. The face is more brightly lit, the background more obviously gloomy. The face less obviously belonging to a solid head. Looking over the top of the lamp, the man is conscious of the rising heat. To this side. To that side. The shadows on the visage are moved dramatically. He is ashamed of this theatre that reminds him of childhood pranks.

So remembering the room he'd occupied at home, he shifts, one by one, the tortoiseshell-backed brushes on the dulled wooden surface of the table. Looks back once more into the mirror's silver mottled centre. Turns his head this way, and that. His looks had not done him much service, yet maybe no disservice.

Face downwards on the bed, he hears his servant ask, "Shall I extinguish the light, sahib?"

At his reply the light goes out in the dour doctor's spare room. Without the lamp's smell the room's atmosphere immediately reconstitutes its mustiness.

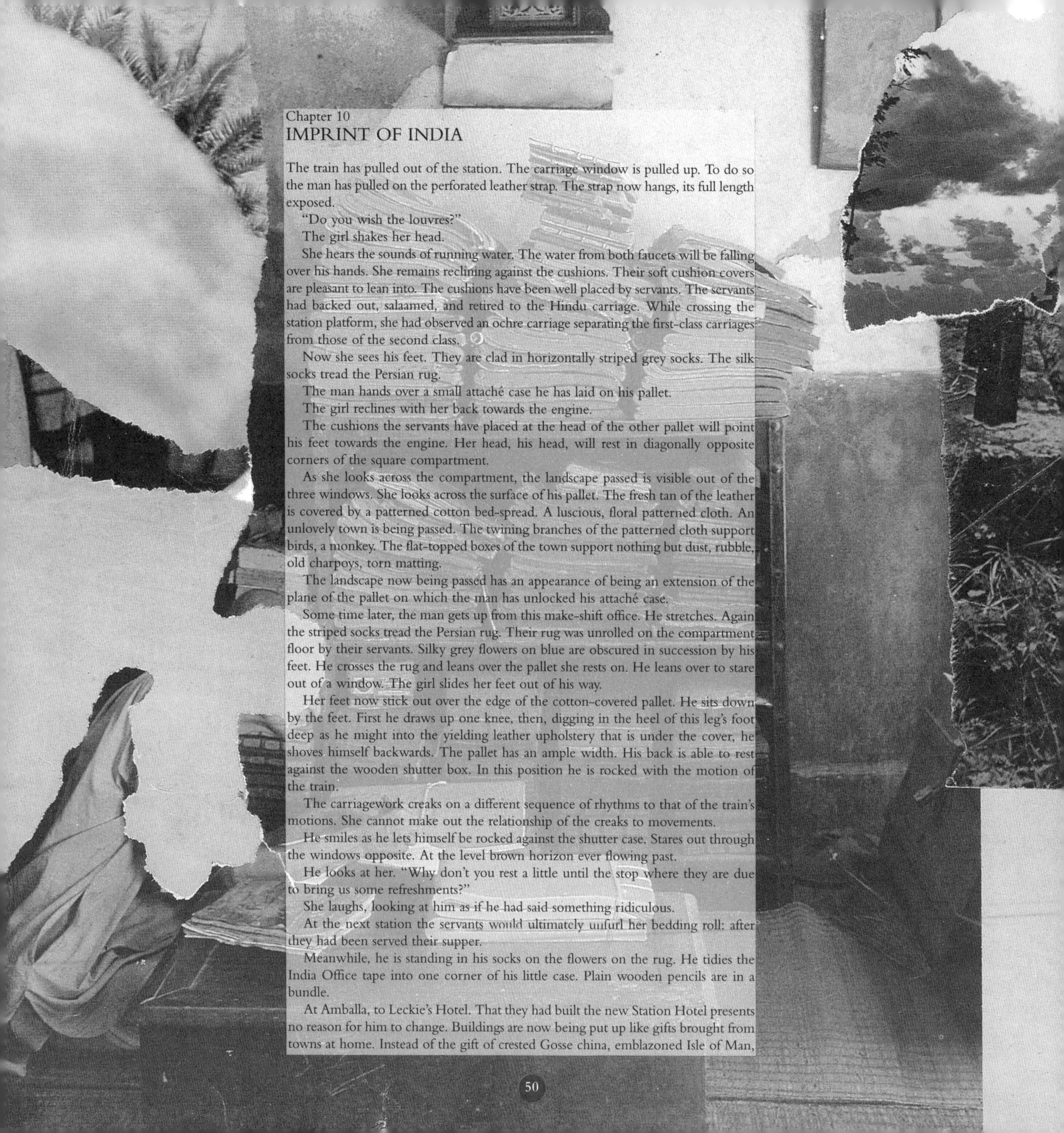

Chapter 10

IMPRINT OF INDIA

The train has pulled out of the station. The carriage window is pulled up. To do so the man has pulled on the perforated leather strap. The strap now hangs, its full length exposed.

"Do you wish the louvres?"

The girl shakes her head.

She hears the sounds of running water. The water from both faucets will be falling over his hands. She remains reclining against the cushions. Their soft cushion covers are pleasant to lean into. The cushions have been well placed by servants. The servants had backed out, salaamed, and retired to the Hindu carriage. While crossing the station platform, she had observed an ochre carriage separating the first-class carriages from those of the second class.

Now she sees his feet. They are clad in horizontally striped grey socks. The silk socks tread the Persian rug.

The man hands over a small attaché case he has laid on his pallet.

The girl reclines with her back towards the engine.

The cushions the servants have placed at the head of the other pallet will point his feet towards the engine. Her head, his head, will rest in diagonally opposite corners of the square compartment.

As she looks across the compartment, the landscape passed is visible out of the three windows. She looks across the surface of his pallet. The fresh tan of the leather is covered by a patterned cotton bed-spread. A luscious, floral patterned cloth. An unlovely town is being passed. The twining branches of the patterned cloth support birds, a monkey. The flat-topped boxes of the town support nothing but dust, rubble, old charpoys, torn matting.

The landscape now being passed has an appearance of being an extension of the plane of the pallet on which the man has unlocked his attaché case.

Some time later, the man gets up from this make-shift office. He stretches. Again the striped socks tread the Persian rug. Their rug was unrolled on the compartment floor by their servants. Silky grey flowers on blue are obscured in succession by his feet. He crosses the rug and leans over the pallet she rests on. He leans over to stare out of a window. The girl slides her feet out of his way.

Her feet now stick out over the edge of the cotton-covered pallet. He sits down by the feet. First he draws up one knee, then, digging in the heel of this leg's foot deep as he might into the yielding leather upholstery that is under the cover, he shoves himself backwards. The pallet has an ample width. His back is able to rest against the wooden shutter box. In this position he is rocked with the motion of the train.

The carriagework creaks on a different sequence of rhythms to that of the train's motions. She cannot make out the relationship of the creaks to movements.

He smiles as he lets himself be rocked against the shutter case. Stares out through the windows opposite. At the level brown horizon ever flowing past.

He looks at her. "Why don't you rest a little until the stop where they are due to bring us some refreshments?"

She laughs, looking at him as if he had said something ridiculous.

At the next station the servants would ultimately unfurl her bedding roll: after they had been served their supper.

Meanwhile, he is standing in his socks on the flowers on the rug. He tidies the India Office tape into one corner of his little case. Plain wooden pencils are in a bundle.

At Amballa, to Leckie's Hotel. That they had built the new Station Hotel presents no reason for him to change. Buildings are now being put up like gifts brought from towns at home. Instead of the gift of crested Gosse china, emblazoned Isle of Man,

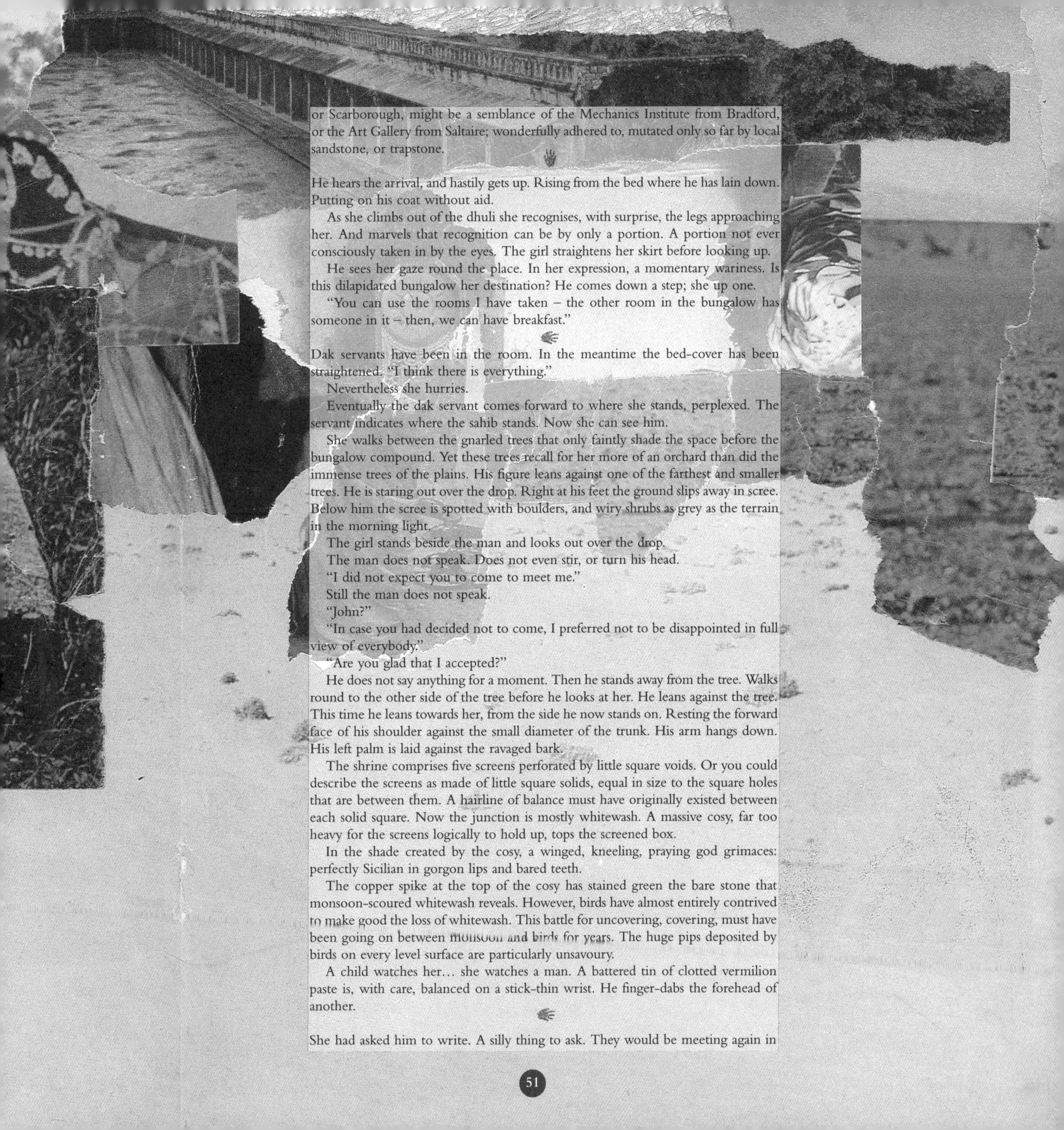

or Scarborough, might be a semblance of the Mechanics Institute from Bradford, or the Art Gallery from Saltaire; wonderfully adhered to, mutated only so far by local sandstone, or trapstone.

He hears the arrival, and hastily gets up. Rising from the bed where he has lain down. Putting on his coat without aid.

As she climbs out of the dhuli she recognises, with surprise, the legs approaching her. And marvels that recognition can be by only a portion. A portion not ever consciously taken in by the eyes. The girl straightens her skirt before looking up.

He sees her gaze round the place. In her expression, a momentary wariness. Is this dilapidated bungalow her destination? He comes down a step; she up one.

"You can use the rooms I have taken – the other room in the bungalow has someone in it – then, we can have breakfast."

Dak servants have been in the room. In the meantime the bed-cover has been straightened. "I think there is everything."

Nevertheless she hurries.

Eventually the dak servant comes forward to where she stands, perplexed. The servant indicates where the sahib stands. Now she can see him.

She walks between the gnarled trees that only faintly shade the space before the bungalow compound. Yet these trees recall for her more of an orchard than did the immense trees of the plains. His figure leans against one of the farthest and smaller trees. He is staring out over the drop. Right at his feet the ground slips away in scree. Below him the scree is spotted with boulders, and wiry shrubs as grey as the terrain in the morning light.

The girl stands beside the man and looks out over the drop.

The man does not speak. Does not even stir, or turn his head.

"I did not expect you to come to meet me."

Still the man does not speak.

"John?"

"In case you had decided not to come, I preferred not to be disappointed in full view of everybody."

"Are you glad that I accepted?"

He does not say anything for a moment. Then he stands away from the tree. Walks round to the other side of the tree before he looks at her. He leans against the tree. This time he leans towards her, from the side he now stands on. Resting the forward face of his shoulder against the small diameter of the trunk. His arm hangs down. His left palm is laid against the ravaged bark.

The shrine comprises five screens perforated by little square voids. Or you could describe the screens as made of little square solids, equal in size to the square holes that are between them. A hairline of balance must have originally existed between each solid square. Now the junction is mostly whitewash. A massive cosy, far too heavy for the screens logically to hold up, tops the screened box.

In the shade created by the cosy, a winged, kneeling, praying god grimaces: perfectly Sicilian in gorgon lips and bared teeth.

The copper spike at the top of the cosy has stained green the bare stone that monsoon-scoured whitewash reveals. However, birds have almost entirely contrived to make good the loss of whitewash. This battle for uncovering, covering, must have been going on between monsoon and birds for years. The huge pips deposited by birds on every level surface are particularly unsavoury.

A child watches her… she watches a man. A battered tin of clotted vermilion paste is, with care, balanced on a stick-thin wrist. He finger-dabs the forehead of another.

She had asked him to write. A silly thing to ask. They would be meeting again in two weeks. Nevertheless, he does as asked. He writes while he waits at the head of the line. Before taking a doolie up country. Only this way will a letter return down the line before he will himself return.

I have to write in the First Class Waiting Room here while they find the cart to take me. I will sleep in a dak bungalow tonight. Let me hope it has not fallen into a state of disrepair. Some are actually abandoned to their fate now that we are at last building some sort of railway system. Only after that scandal of the last famine. Nothing yet has happened for me to write about. I am incredibly lazy in transit. I sleep, eat, and read. One cannot imagine what else to do seeing the engine is doing all the work for one. Fifteen hours on the train and I feel I am still in some sort of motion, even the pen seems to sound like the engine. I take it you can read my writing. Most people have no difficulty.

Regards

Jonathan

He cannot think how else to sign the note. He wishes he could express more personal thoughts with ease. He has long since lost the habit in letters home. Here in India, letters might be handed round. Such is the custom. Correspondence helps to keep a sense of national identity; as well as pass the time.

Talk concerns the equalisation of the sugar dues, the Roman alphabet, steam navigation, the abolition of the transit duties… The usual desultory shop-talk, as he joins the gentlemen already assembled.

"Well," someone says, "This is new for you surely, Jonathan. A nocturnal tête-à-tête with a young lady."

Nothing was ever missed in these cantonments, nothing!

"We went to hear the frogs," he says evenly. His policy is always to tell the truth… But then it dawns on him how out of his mind this must make him seem. With some difficulty he keeps a sober face. His audience certainly seems nonplussed. Whether out of politeness, or suspicion, or disappointment, none of them laughs.

"She's an odd girl," comments the doctor.

The doctor hands the girl one sheet of a letter.

She takes it. A handwriting she does not know. A last sheet. It is signed, "Nathan"? Her eye scans the sheet. "… my compliments to Miss Urquhart"… to the starting of the paragraph.

How is the work? I hope you are not wearing yourself out. My compliments to Miss Urquhart. I envy you; my rooms sound very empty. The same five men at the end of every gruelling week offer nothing by way of a change, or relaxation.

Regards

Nathan

The girl reads the letter again; slowly. Then there seems nothing for it but to hand the sheet back.

The doctor glances up sufficiently to watch it placed on his desk. The doctor returns his concentration to his cup, to his stirring the tea until it will be cool enough to drink. The servants never remember that one's tongue and gullet are not made of leather. Nettie had tried in the early days to train them… the doctor thinks of those days as he stirs. And as he always does think this way as he stirs, he does not mind stirring: something of a ritual. After a while, he says, "It seems it is not all one- sided, this day-dreaming. What are you going to do about it?"

AFTER-FIND

"There are hours which one can never forget, into which the enjoyment generally spread over large portions of life seems as it were concentrated; among these there are few more happy than those in which we realise another climate, the air, soil and vegetation being totally different, and all inspiring new and delicious sensations; when a new page of the endless variety of creation lies open before us."

W.H. Bartlett, *The Nile Boat*